The Concept of Object
as the Foundation of Physics

San Francisco State University Series in Philosophy

Anatole Anton
General Editor

Vol. 6

PETER LANG
New York • Washington, D.C./Baltimore • San Francisco
Bern • Frankfurt am Main • Berlin • Vienna • Paris

Irving Stein

The Concept of Object as the Foundation of Physics

with forewords by

C.W. Kilmister &
H. Pierre Noyes

PETER LANG
New York • Washington, D.C./Baltimore • San Francisco
Bern • Frankfurt am Main • Berlin • Vienna • Paris

Library of Congress Cataloging-in-Publication Data

Stein, Irving
The concept of object as the foundation of physics/ Irving Stein.
p.   cm. — (San Francisco State University series
in philosophy; vol. 6)
Includes bibliographical references.
1. Quantum theory. 2. Physics—Philosophy. 3. Object (Philosophy).
I. Title. II. Series.
QC174.12.S84   530.1—dc20   95-24071
ISBN 0-8204-2537-0
ISSN 1067-0017

Die Deutsche Bibliothek-CIP-Einheitsaufnahme

Stein, Irving:
The concept of object as the foundation of physics/
Irving Stein.- New York; Washington, D.C./Baltimore;
San Francisco; Bern; Frankfurt am Main; Berlin;
Vienna; Paris: Lang.
(San Francisco State University series in philosophy; Vol. 6)
ISBN 0-8204-2537-0
NE: San Francisco State University: San Francisco State...

∞

"Nothing is Neutral:
In Every Paradigm
Is An Ideology."

"Good Philosophy—
and Better Physics—
is the Penetration
of Ideologies."

Acknowledgements

My deepest thanks to H. Pierre Noyes, Stanford Linear Accelerator Center, Stanford, Ca.; to V. A. Karmanov, Lebedev Physical Institute, Moscow, Russia; and to Anatole Anton, San Francisco State University, San Francisco, Ca.

THE SAN FRANCISCO STATE UNIVERSITY
SERIES IN PHILOSOPHY

This series is designed to encourage philosophers and scientists to explore new directions in research, particularly directions that may lead to a re-integration of philosophy with the sciences, the arts and/or the humanities.

The series is guided by three premises.

1. The intellectual division of labor into distinct academic disciplines is a product of changing historical circumstances and conditions (including developments within the disciplines themselves).
2. The current intellectual division of labor has outlived its usefulness in many ways.
3. There is a pressing need to re-integrate the metaphysical and evaluative concerns of philosophy with current work in the sciences and their associated technologies, the humanities and the arts.

Works in this series are intended to challenge social and philosophical preconceptions that block the re-integration of philosophy with other disciplines, and at the same time to maintain unquestionably high standards of scholarship. The third volume in our series, *The Concept of Object As the Foundation of Physics* by Irving Stein, exemplifies our intentions. Neither philosophers nor physicists have succeeded at the task of providing an ontological foundation for modern physics as a consistent whole. Irving Stein has provided powerful considerations for the view that future progress in our understanding of physical theory no longer can rest on the notion that the concepts of space, time, and object are foundational or even that the concept of physical law is fundamental. Stein's work is aimed at showing the feasibility of providing an ontological foundation for the whole of modern physics. What he has done in the essay which we publish here is to develop in detail such a foundation for a one-dimensional non-interactive physics and lay the basis for its extension into interactive physics. From this foundation he has been able to derive the concepts of space, time, and object as they appear in modern physics. Furthermore, a lesson of his research shows the need for continued work in what may be called critical physical theory. If Stein is correct, the disciplinary boundaries between philosophy and physics are artificial and exist to the detriment of both fields.

Anatole Anton, *General Editor*
San Francisco State University
Department of Philosophy
1600 Holloway Avenue
San Francisco, CA 94132

Foreword

by C. W. Kilmister

Physics has become so very difficult to understand that a strange view, that the concept of understanding needs to be re-defined so as to be relevant to physics, is now widespread. The symptom of the difficulty is the continuing failure to find a common ground between relativity (gravitation) and quantum mechanics. It is becoming more and more likely that this failure is not just a technical matter, but is simply caused by the failure of understanding.

How can one break out of this cycle? This book presents one man's quest. Irving Stein starts from an analysis of the notion of object, and what is characteristic of his book is its complete integrity. Step-by-step he peels off the onion-skins obscuring the understanding of, first, the classical object in space and time, then the quantum object. But still paradoxical features remain, so he is finally led to a deeper formulation of his own, which presents a challenge to the reader: accept or do better!

Kings College, London (Ret.)
February, 1995

Foreword

by H. Pierre Noyes

Most physicists would agree that the greatest advances made in physics during the twentieth century are the Einstein theory of relativity and quantum mechanics. Most laymen interested in these subjects often find them difficult or downright paradoxical. Yet few physicists have found these facts sufficiently challenging to force them to entertain the suspicion that their technical expertise might need a firmer foundation. They do not even take umbrage when a distinguished physicist and historian of science (Schweber) asserts unequivocally that "How to synthesize the quantum theory with the theory of special relativity was—*and has remained*—the basic problem confronting 'elementary' particle theorists since 1925-27." One suspects that this thorny problem seems less exciting to them than areas where they can exhibit technical virtuosity.

Irving Stein is that rare physicist who has insisted on understanding for himself what could serve as a firm *conceptual* foundation for modern physics. Early in his career he saw a particular way to combine relativity and quantum mechanics in an intimate union, but a union that he subsequently realized raised problems of its own. This book is a description of his odyssey, searching at each point for consistency, and finding once again paradox or incompleteness. The foundation which emerges is still to be done when it comes to describing interactions, but does resolve in his terms what has to be meant by a "quantum object." Irv and I were once fortunate enough to go over Stein's derivation of the Schroedinger Equation with John Bell, and to convince John that it is valid in context. Unfortunately, John's untimely death prevented us from discussing Stein's derivation of the Dirac equation with him. The reader can rest assured that, despite the unfashionable concern with ontology which Irv finds fundamental, the results are technically valid and quite unique.

The book develops historically, which enables the reader to share both the excitement which Irv felt as one point after another came within his grasp, and the frustration he suffered when this stage of the journey once again proved to be a halting rather than a resting place. I hope you will enjoy this intellectual adventure as much as I have.

SLAC, Stanford
February, 1995

Abstract

I attempt in this work to give meaning to and analyze the concept of a simple, non-structured object. First, the concept of a classical object is analyzed and found to be inherently deterministic, where determinism is nothing but mathematical analyticity, a rather strong but necessary restriction on the objects required for classical physics. Yet, it is found that even such a strong restriction as analyticity is not sufficient to define an object consistently. In order to remove this inadequacy I consider an object as a classical random walk; this leads to the amazing consequence that in the limit as the step length approaches zero, the theory of (special) relativity results. Further analysis of the concept of the random walk object allows us to integrate the concept of mass into that of the concept of object; it also leads to a pre-quantum theory de Broglie relationship. In order to give up what is seen as utter confusion between a single object and an ensemble of objects in the random walk concept, the concept of a non-preferential random walk for a *single object*, which then defines a quantum object is introduced. Such an object is the basis of quantum mechanics from which the Schroedinger equation is derived. Further development of this idea then leads to the Dirac equation.

The nature of measurement is investigated and found to be distinct from that of interaction and fundamental to understanding the ontology of physical reality. Although it is the purpose of this work only to give a reasonable, coherent definition of the concept of object, *it will be seen that the theories of relativity and quantum physics arise out of the unfolding of the concept of object presented here.* In the end, the concept of object itself is found not to be absolutely basic and dissolves into the concept of what I call *nonspace*, which is found to be the fundamental ontology. Objects arise from restrictions on nonspace and classically deterministic objects result from further restrictions on the concept of nonspace.[1]

Contents

Preface

Why is it that the speed of every object in the universe must be less than that of light, and that of light is the same in all galilean frames of refererence?

If an elementary particle passes through a grating, or, equivalently, if it passes through two or more slits in a diaphragm, or for that matter, *any* object does, elementary or not, its wave function changes as it goes through the slits. This is despite the fact that there generally is no interaction—no interaction potential—between the grating and the particle. The question then is: if there is no interaction, why is there a change in the wave function? Furthermore, the particle passes through all of the slits without dividing. How can such a process be understood?

What then is the nature of an object, that it can pass through slits without dividing and whose kinematical state, condition, or wave function is transformed without interaction; and furthermore whose speed is always less than that of light? It is one of the purposes of this work to answer such questions by establishing an ontology for physics, a reality "ursprung."

Such a work is a work in physics, but it must be informed by a philosophical sensitivity.

Oakland, CA   I. S.
February, 1995

I

Prologue

Many years ago, when I was the most junior faculty member in the physics department of a medium-sized midwestern college, during an evening social function I asked two of the more senior—although still then young—faculty members if they understood quantum mechanics. Since I had struggled in vain to understand and since neither written text nor my graduate advisor, who was quite famous, seemed to understand, at least to my satisfaction, I was more curious than hopeful. The first physicist, who already had quite a reputation as a theorist said, very earnestly, while looking down at the floor and nervously smoking a cigarette (I remember vividly) that "reality is all in the mind anyway, so what difference does it make?" The second physicist, who later became the associate director of a large midwestern research facility stood straight up, bald, very erect, and said, very quietly, that understanding is just a matter of getting used to what is being said. (I remember his standing up so straight because later on in the evening he said that homo sapiens sapiens was not supposed to be erect and the fact that it was was what caused our back problems.) Neither of these answers satisfied me. Even if it was all in my head, there were many things in my head that I *could* understand; I could understand Newtonian physics and I could understand relativity (I thought then). As for the answer of the second physicist, I knew I was used to many things that I didn't understand, quantum mechanics for one, and that some things I was not used to but did understand, such as money.

What was my problem? Did I not believe that quantum theory was correct? No, I had not and never did have the slightest doubt that not only was it correct but that it was the most powerful theory ever invented, and that it and relativity would be the basis for all future developments in physics. Did I feel, like Einstein, that it was an ad hoc theory, to be put on a firm basis later on? I did not feel that it was ad hoc, but I thought that something, I knew not what, was missing, that it was incorrectly formulated and that its correct formulation was inhibited by a particular philosophical outlook—a rather incoherent empiricism—to which we were—and still are—all subject. Specifically, I could not accept the role to which measurement was assigned in quantum mechanics; if measurement caused an "uncontrollable" change in what was being measured, then what was the measurement measuring? I could not accept the Uncertainty Principle as the "essence" of quantum mechanics, for then how could there be *independent* physical realities, as is claimed, such as position and momentum, and energy and time that were not simultaneously measurable? Furthermore, the fact that an object, such as an electron, proton, neutron, etc. could go through two or more slits *without dividing*, or that an object could go from one point to another without having a path, was indeed mind-boggling, even though not in doubt. It was simply that I could not understand how there could be laws of physics, or any kind of laws, without determinism. Statistical laws are not laws; they are either based on lack of knowledge or, as in the case of quantum mechanics, based on a lack of causality or determinism. Therefore, since I knew that quantum mechanics was correct, I concluded that it must be incorrectly interpreted due to the positivistic fog we were all in. And I was resolved to get us all out of it. But all of my efforts to do so failed miserably; the difficulty in understanding was not a matter of interpretation, or of semantics, or of meaning, I soon realized, and also not a matter of the theory, the equations being wrong; rather it was a matter of incompleteness, that is, there was no clearly stated *ontology*. That is, our understanding of the world discovered by quantum mechanics was expressed in terms of measurement and not in terms of the reality that gives rise to the results of the measurements. Because we were either unable or unwilling to pursue an ontology, we were fated to remain with the legacy of our previous and very successful ontology, that of determinism. The tragedy of this, as I will show, is that although most of us have consistently denied a deterministic ontology, nevertheless the strangeness and incomprehensibility we find in quantum mechanics is only in response to deterministic expectations. The only statement of a quantum physics based on a deterministic ontology, that by Bohm, Vigier, et al., which, in my opinion, has failed, itself in recent years seems to have degenerated into incomprehensibility.

Much interpretative work has been done since I first posed these questions to myself. Very subtle and dramatic experiments have been performed to further test (the already proven) quantum mechanical concepts—all tests confirmed these concepts, of course—but the theory is still in its original disarray. Thus, I was further encouraged in my belief that the work that had to be done was not interpretive, but conceptual; that is, it was required to develop quantum mechanics based on concepts more fundamental than those then being used. I came to the conclusion that such a theory had to be based on the concept of *object*. That is, the results of *all* physics—as it eventually turned out, not only quantum but also classical and relativistic—must arise from the nature of the object itself (at least for non-interacting objects). Thus, I began to look for an object *ontology*; that is, the very *nature* of objects from which all of physics would flow. Newton had an ontology; he very clearly said that every object in the universe has as its fundamental property, inertial mass; he stated the existence and independent realities of space and time. He furthermore stated that every object in the universe also has the fundamental property of gravitational mass and that both of these were equal. That is, he made statements about *objects* and their *interactions*. To him, the laws of physics were the space-time behavior of objects either free or subject to interactions with other objects. Even if it was exoteric, in that the nature of an object, as mass at a given location at a given time, was defined and expressible largely not of intrinsic behavior, but rather in terms of response to each other, nevertheless Newton did have an ontology. This behavior, expressed in his 2nd and 3rd laws of motion, exoteric or not, has nothing to do with measurement, observation, or any other kind of *epistemological* consideration. In quantum mechanics, on the other hand, we are unable to state our theory in terms of objects, or, for that matter, in any kind of ontology. To quote Schiff from his well-known textbook, *Quantum Mechanics*, "the number of measurements that result ... is proportional to the square of the magnitude of the coefficient of u_ω." He then says: "This enables us to associate a probability function with any dynamical variable."[2] That is, an object's state, which now defines the condition of the object, is defined by stating what the probability distribution of some dynamical variable *would* be *if* the appropriate measurements were to be made on an *ensemble* of objects (meaning, all having the same state function). Amazing! In order to know the state of the object, it is necessary to remove it and an entire ensemble (all its relatives?) from their states. Amazing! Go fight Pupin Hall.

While doing my dissertation, I remember asking my thesis advisor, a very well-known and distinguished theoretical nuclear physicist, three questions:

1. If an object approaches a 2-slit diaphragm and doesn't hit it, it then must go through the two slits to the other side. Thus, even though there is no interaction, there is a change in the wave function. What *caused* the change in wave function? If interactions are not the cause, what else is there that is the cause?
2. How can one measure mass? If it is attempted to measure mass gravitationally—say, by the stretching of a spring—then, since both the position x, and the velocity v, must be known, the Uncertainty Principle is contradicted. If one attempts to do it by the inertial property of mass using $F = dp / dt$, then once again the Uncertainty Principle is contradicted, since change in momentum, and thus momentum, must be determined at an instant of time or in a very short time interval.
3. Since $E = mc^2$ and $(\Delta E)(\Delta t) \sim \hbar$, how can m be determined in a finite time?

There was no response.

From then on I knew that I would understand quantum mechanics only if I knew *what* an object is, an object being the conceptual basis of the idea of particles, of particles "stripped down," denuded of all but kinematical and mass properties.

Later on I realized that it was not only quantum mechanics but relativistic and even classical physics that also required the concept of object in order to be understood. That is, I asked myself, what must be the nature of an object to give rise, in classical physics, to determinism; in relativistic physics, to a maximum velocity; in quantum physics, to the Schroedinger and Dirac equations?—respectively, classical, relativistic, and quantum objects, the latter being further subdivided into Schroedinger and Dirac objects. Each of these different physics should, I felt, arise from the very nature of its own kind of object. (The further extension of the ontology into the nature of interactions, giving rise to the concept of field, is not done here—interaction indeed is a very difficult concept to understand.) Thus, for me at least, the ontology of physics is an ontology of objects. I conceive an object as the source, the "ursprung," from which arises the appropriate physics. In summary then, in my view, in order to develop a coherent theory of quantum mechanics, it is required to state its ontology; that is, not just the results of measurement, but what *is*. The work done here claims that the eventual ontology arises out of an ontology of objects. However, in order to develop an ontology of objects for quantum mechanics it is necessary to first clarify and develop the concept of object for both classical and relativistic physics, since the concept of quantum object is based on the concepts of the relativistic and classical objects. Thus, the original query, "Do you understand quantum mechanics?" led to the more general question, "Do we understand (non-interactive) physics?" My answer is yes only if we understand its ontology, in this case the concept of object. And to understand the concept of object means to be able to define it, to give it meaning. But on what basis can meaning be given to so basic a concept as that of object? A good deal of the source of such meaning will be found in the idea of "non-preference;" i.e., unless one alternative is inherently preferable to another, neither alternative can exist. This idea of "non-preference," however, is not always as straightforward as it might seem—it is always non-preference within a context. Thus, in a collection of random walk objects where the step length approaches zero, and there is equal preference for a left or right step, the distribution of locations is a gaussian distribution. On the other hand, if, as in classical physics, it is not a random walk, but simply a collection of objects making walks analytic in time, then we would expect the distribution of locations to be flat.

The first clue as to what to do to reformulate quantum mechanics was the realization that the only two possible values of the velocity operator (the eigenvalues) in relativistic quantum mechanics are ±c. This, of course, must be sheer and utter nonsense. How could the only possible speed of *all* objects be c? Relativity theory clearly states that *all* objects of larger than zero mass have speeds less than c. The blindness of physicists (with due humility, even the attempts of Dirac, Schiff, et al) to the ordinary courtesies due the law of contradiction jarred me; however, it is certainly "justified" if the theory is wonderful and it is only logic and not experience that is confounded, which is the case. (Dirac, being Dirac of course, knew the correct answer; that is, that the velocities "observed" are "average" velocities. But as far as I know, he never resolved the apparent contradiction with relativity. See Dirac, *Quantum Mechanics*, 1947, 3rd ed., p. 261.) Furthermore, the greatest physicists had usefully violated logic before, Einstein with the photon concept, in apparent contradiction to the wave theory of light, Bohr with his unavoidably but purposely botched logic of stable nonradiative electron orbits in apparent contradiction to electromagnetic theory, and Newton with his absolute frames of reference "contradicting" his relative frames of reference. All three theories, however, are apogees of physical theorizing. Nevertheless, since I was concerned with understanding the theories of physics, particularly quantum mechanics, which was to me rife with irrationality, I wanted to resolve its incoherence, even though the theory is wonderful; I thought that I could do this by making quantum theory a theory of objects.

Since irrationality seemed to be such a productive force for the nurturing of the flowers of physics, I switched; I decided that perhaps, if treated gingerly, irrationality could also be the root of rationality (hoping, of course, for the eventual dialectic resolution of the contradiction). I thus decided to take the path I found out later others had also taken; that is, I assumed every object moves in random discrete steps, equal distances in equal times, always at the same speed, c. In fact, I went even further than this; I defined objects in terms of random walks, and thus distinguished each object from every other by its particular random walk. This is the mother of all that follows, although here the children are cleverer than the parents.

Two consequences immediately followed from defining the object as a random walk: first, I found that the formula for the standard deviation of a random walk distance distribution simulates the length contraction formula of relativity; and second, an uncertainty principle expressed as the product of the standard deviations of position and "momentum" of the random walk distribution appeared which could be made equal to Planck's constant. I was fooled; I thought that I had, so to speak, with one stroke, discovered that a random walk leads to both relativity and quantum mechanics. No, not yet.

Later on I realized that a random walk, based on the equality of the magnitudes of space and time steps and therefore, as the step lengths approached zero, defining at a given instant of time and at the same position both instantaneous and average velocities, led directly to the theory of relativity. *That is, I discovered that it is not the space-time transformation property that is fundamental to the theory of relativity but rather the above mentioned space and time identity of the random walk steps.* This result by itself is worth the price of admission, even though it does not yet give us quantum mechanics. But this concept of the identity of space and time in the small, which was found to be the source of relativity, is also, surprisingly, one basis upon which quantum mechanics seems to be built. *That is, quantum mechanics, even non-relativistic quantum mechanics, is based on a relativistic concept*, or more accurately, on the concept that also gives rise to relativity.

However, the big thing that distinguishes quantum mechanics from both Newtonian and relativistic physics arises from the fact that in a (non-preferential) random walk there is no reason for an object to go in one direction instead of the other. Therefore, I conclude, an object must go in *both* (one dimension) directions and thus *at any given instant of time exists at more than a single location* (in the absence of a "measurement"). This very peculiar result is at the heart of quantum mechanics;

that is, the statement that the random walk of an object is directionally non-preferential is the essential statement of quantum mechanics. From this it can be shown that the state of an object is defined by the values of *its* property; e.g., positions, and not through the property of an *ensemble* of objects.

Furthermore, the frog of measurement, so contemptuously dismissed by me earlier had, in the end, turned out to be the prince of transformation; if not quite the ruler of the kingdom, certainly the chief parliamentarian of basic concepts. Between measurements, an object does not exist in space; it exists at many locations at the same time, that is, it exists in what I call *nonspace*. Such a tragic demise of the reign of space as fundamental, an idea upon which so many geometers and classical physicists, to say nothing of space-laced philosophers, made a comfortable living, is really not to be regretted; we have been stuck too long in the reality cliches of outmoded ways of thinking, both scientific and philosophic.

The introduction of the concept of nonspace and the concomitant understanding that a random walk in nonspace takes only a single *spatial* step between measurements leads directly to the Schroedinger equation of a free particle (object) in one dimension. It is in this way that measurement came back into its own ontologically, no longer as a magic wand but rather as a "reality probe;" that space, except at moments of measurement, disappeared and reappeared as nonspace; that time became, again except at the moments of measurement, imaginary (in the mathematical sense).

The further realization that the object does take "steps" in nonspace in between measurements and, as will be shown, in imaginary time, although these steps are not of fixed but of varying length, then leads, with the help of calculations suggested by Feynman and done by others, directly to the Dirac equation. This modification of the simple random walk, the step lengths no longer being fixed, but having a continuous range with an average length equal to the former fixed length, which can be identified with the Compton wavelength, turns out to be the key to the derivation of the Dirac equation. That is the reality from which the properties of the Dirac object are derived. *That* is the ontology, *that* is the nature of physical reality in my theory.

Such a radical restructuring of the bases of physics, no longer space, time and, as we will see, even object, but rather nonspace and imaginary time from which, by measurements, objects may arise, is less an imaginative assay into reality than the logical consequences of a demand for an ontology. It is for such reasons that I give up the idea that space, time, and object are fundamental, replacing it with an alternative conception. It is this way that the theory has developed: what are the natures of objects so that the appropriate physics arises directly from them? In fact it turns out that only the final object, the Dirac object, is even *possibly* logically coherent, the others only being way stations. Most significantly, the logical point in this work is to discover the nature of an object, its definition and its meaning, and not that of some physical theory. It is apparently only *incidental* that the correct physics arises, whether it be classical, relativistic, or quantum, almost as logical afterthoughts, but indeed, gratifyingly confirming ones. Only then can we see that even the concept of object itself is not the most basic reality, since the concept of a quantum object leads to the more fundamental reality of nonspace and imaginary time.

II

Introduction

All experiments and observations in physics are made on objects, where by "objects" is meant the elementary particles and aggregates of them. All experiments and observations, of course, are based on measurements. Measurements are performed by classical objects, which are sufficiently massive single objects, or by aggregates of elementary particles. Such classical objects occupy a given position or positions in space at a given time. In this work, I ask the questions: What is the nature of an object so that it can give rise to mass, space, and time? What must be the nature of an object, classical or otherwise, so that it can give rise to such fundamental phenomena of physics as relativistic and quantum physics, to say nothing of classical physics? What is the nature of an object such that it can have a maximum velocity and, if it is an elementary particle, can go through two or more slits without dividing? That is, what is the nature of an object such that it can give rise to the theory of special relativity expressed in the Lorentz transformation and quantum mechanics expressed in the Schroedinger and Dirac equations? The work presented here claims to have discovered the basic nature of an object from which the above "laws" of physics can be derived. But it furthermore shows that there is a reality more fundamental than that of object, one, in fact, from which objects themselves may arise if measurements are made.

In order to understand the nature of an object in its most coherent formulation, it is first necessary to understand the nature of a *classical object*, that is, a more familiar concept of object, one that is a mass "associated" with a space function of time. That is, in classical physics, mass, space, and time are the fundamental realities and an object is defined in terms of them. How is this done?

Space and time in classical, Newtonian physics are independent realities; there is no correlation between values of space and those of time. A classical object is defined as that which does correlate values of space and time. Furthermore, a classical object is a mass and not just any function but rather an *analytic* space function of time; that is, a function that is "deeply continuous." Since the functions are analytic, the physics must be a deterministic physics.

Furthermore, I claim in this work that space and time exist *only* where and when objects exist. That is, space and time values are no longer independent of each other but exist and are defined by the functions defining the objects. The justification for such a claim is that since all measurements are made on objects, the claim for the existence of space and time outside objects must be made on grounds other than measurement: I know of no such grounds. On the other hand, as I show, the results of quantum mechanics actually *prohibit* the existence of space and time outside of objects.

Further consideration then shows that the definition above of the concept of object as a space analytic function of time is basically flawed; that is, since no analytic function is to be "preferred" to any other in defining an object, the velocities of the objects must be equally distributed; that is, as many objects can have one velocity as any other other velocity. Therefore, since the range of velocities is $(-\infty, \infty)$, the probability of an object being in any finite region of space for any time interval, $\Delta t > 0$, is zero. Furthermore, the concept of the *mass* of an object is left dangling; the best that can be said about the relationship of mass to object is that it is "associated" with a space-time function. In order to remove these difficulties, especially the first one, it is necessary to define an object by a relationship between space and time *more* restrictive than the analytic one. This is done by defining the object function as a space random walk in time where both the space and time steps are everywhere of equal magnitude. Being everywhere of equal magnitude means that except for the space step being positive or negative while the time step is always positive, space and time are identical, indistinguishable, the same kind of reality. Therefore, the ratio of the space step to the time step, the "instantaneous" speed, is always constant. *Such an identification is the basis for the special theory of relativity.* Thus no further postulate such as the velocity of light being constant or that there be a maximum velocity to all objects is required. It is only *then* that the concept of (relative) velocity and velocity frames of reference are defined for random walk objects, thus producing an entirely different ontology of physics than in the usual relativistic physics. The Lorentz transformation then immediately follows.

I then define the mass of an object as inversely proportional to the magnitude of the space-time step; although this identification is not successful since it is necessary to let the step length approach zero, it is the conceptual basis for the eventual successful incorporation of the concept of mass into that of object.

Further investigation shows, however, that a classical random walk is a *preferential* walk, for which there can be no justification. In order to remove this defect in the theory, it will be shown to be necessary to allow an object to occupy many (two or more) positions at a given instant of time. Such radical surgery to the concept of space is the *transition from classical to quantum physics.* From this concept of a multi-position object and, as will be shown, existing in imaginary time, I then can derive the Schroedinger equation. With a further modification of the concept of object, the Dirac equation follows. *It is the object being many-positioned, at an instant of imaginary time, which is the ontologic basis of relativistic quantum mechanics.*

However, even though I start out attempting to find an adequate definition for a *single* object, I end up with the Dirac theory, which is a physics of many objects. Thus, it is not possible to identify nonspace with just a single object; many objects might "arise" from nonspace. Therefore, it is no longer the *object* that is the ontological basis for physics, but nonspace and imaginary time, which turns out to be the ontological basis of the object itself.

How *do* objects "arise" from nonspace? They are not *in* nonspace (any more than electrons are in the nucleus, although they arise from the nucleus in radioactive decay). Objects arise from nonspace by *measurements* on nonspace.

Then, if nonspace can no longer be identified with an object, what *is* nonspace? A general answer to this question is: nonspace is "all possibilities," *where the meaning of "all possibilities" is defined in terms of classical objects and measurements*. Thus, we have the interesting situation where quantum concepts and, thus, basic reality, must be defined in terms of classical concepts, such as measurement, although the classical concepts are derivative from the quantum concepts. Therefore, although all of reality must be *defined* in terms of measurements, which is a classical reality, all of reality is not *only* classical reality. Far from it. In fact, classical reality is only a highly specialized case of the more basic, quantum reality, resulting from *restrictions* on the nonspace (measurement, interactions, numbers of objects, dimensionality, and many more) from which classical (massive) objects arise. How and why restrictions on the "all possibilities" nonspace arise is not known. But it is these restrictions which eventually produce, among other results, the classical world. It is because of measurements, a classical concept, that the discovery of nonspace, a non-classical concept, is made. The basic ontological problem remaining is to discover the sources of the many restrictions on the nonspace.

What is claimed in this work is that an ontology has been laid out for physics, at least for a one-dimensional, non-interacting physics. *By "ontology" is meant the origin in reality of the results of measurements*. This is done not by starting out with a set of hypotheses or axioms, but by attempting to define the most basic concepts that we have in discovering the world around us, namely those of object and measurement.

The work presented herein is a working out of these ideas and inspirations, and shows the possibility of one consistent ontology for all of physics, whether classical or quantum. It is organized into nine further chapters, each building upon the preceding.

In chapter III the concept of the classical object is developed. Such a concept is, at first, a rather primitive one, arising out of more or less immediate experiences which give rise to theories such as the kinetic theories of gases or simple non-structured objects. The basic ideas on which this next chapter, chapter III, is built are the following:

1. Objects are the basis of all (physical) reality.
2. A *classical object* is defined as a given space function of time associated with a mass.
3. The space-time function defining an object is eventually determined to be a real analytic function over the entire time domain.
4. Space and time are reasoned to be properties of an object rather than independent realities. Thus, contrary to Newton's ideas of absolute space and time, I give reasons to support the argument that space and time exist only where and when objects exist. (Newton, of course, did admit the concept of relative space and time; that is, the space and time of particles relative to each other, but thought that they presupposed the existence of absolute space and time.)

In chapter IV we see that although a basis for the concept of a classical object was established in chapter III, the concept is still incoherent. Modifying the concept of classical object to overcome this incoherence gives rise, amazingly, to the special theory of relativity. It is then required to modify and further develop the concept of object by defining anew its space-time framework.

1. A *relativistic object* is defined, not just by an analytic function as is a classical object, but also by the limit of a random walk as the step length approaches zero. However, just as for a classical object, the mass exists as a separate property of the object. Later on in this chapter it is seen that the need for such a property as that of mass is required by the possibility of interaction.
2. The step length and the step time are defined as identical in magnitude and differing only in that the step time is positive and the step length is either positive or negative. Thus, the ratio of the two is always only a single magnitude, $c = \ell / \tau$.
3. It is assumed that the limit of the ratio as $\ell \to 0$ exists and has the same value, $c = \ell / \tau$. That is, the instantaneous velocity of an object is always $\pm c$. The fact that the eigen-

values of the velocity operator in relativistic quantum mechanics are ±c is now seen to follow directly from the nature of the relativistic object.

4. Thus, at any instant of time, two ratios are defined by such an object: i. The *instantaneous velocity*, whose value is always ±c, and ii. The instantaneous *average* velocity, $\bar{v}$, whose magnitude must be always less than or equal to c; that is, $\bar{v} \leq c$. Thus, although the instantaneous velocity is everywhere discontinuous, the limit of the average velocity; i.e., the instantaneous average velocity can be defined everywhere as analytic.

5. The concept of the velocity of an object *relative* to that of any other object is then defined, giving rise to the concept of *velocity frames of reference*.

6. From this definition of an object and frames of reference the space-time Lorentz transformation is derived.

In chapter V, in order to prepare for the specification of a quantum object, I analyze the (classical) random walk object.

1. This object is defined, as is the relativistic object, as a random walk, but one whose step length is larger than zero; i.e., $\ell > 0$.

2. The mass of the object, m, is defined by ℓ. Thus, m does not define an extra or other property of an object; rather it defines or is defined by the step length of the object. With this identification, interaction, as in Newtonian or relativistic physics, is no longer required as the basis for the property of mass. Nevertheless, interaction among objects requires the existence of mass.

3. Now, assuming that the average velocity, $\bar{v}$, of such an object exists, the space-time function of such an object is simply a random walk. An ensemble of such objects (i.e., having the same $\bar{v}$) then, for any n, has a binomial distribution of positions and momenta, and the product of the standard deviations of position and momenta is found to be

$$(\Delta x)\,(\Delta p) = \frac{\hbar}{2}, \qquad \hbar \text{ a constant}$$

where the value of the mass, m, for any given ℓ has been defined as $m = \hbar / 2\,\ell c$, an uncertainty relationship remarkably like, yet quite distinct from, the Heisenberg Uncertainty Principle,

$$(\Delta x)\,(\Delta p) \geq \frac{\hbar}{2}$$

If m is defined this way then the step length, ℓ, is seen to be nothing but the Compton wavelength, λ_c. We see that these results have nothing to do quantum theory but arise from the (classical) random walk object only; i.e., these results essentially express the mathematical concept of randomness and not a physical theory such as quantum mechanics.

4. It is clear, however, that as long as $\ell > 0$, $\bar{v}$ cannot be defined in a finite time as it can be for classical and relativistic objects; on the other hand if $\ell = 0$, then $m = \infty$. Furthermore, for a binomial distribution, the standard distributions for x and p are

$$\Delta x = \ell\,\sqrt{n} = c\,\sqrt{\tau}\,\sqrt{t}; \qquad \Delta p = \frac{mc}{\sqrt{n}} = \frac{mc\,\sqrt{\tau}}{\sqrt{t}},$$

while for quantum mechanics Δx increases proportionally with time and Δp is independent of time (as long as no measurement is made).

In chapter VI the Schroedinger object is defined and the Schroedinger equation is derived.

1. This is done by removing the last vestige of classical mechanics, the assumed existence of space between measurement locations. That is, if there is no preferential direction for the object to move, it must move in both directions and thus will exist simultaneously in more than a single location; I call such a set of points a *nonspace*. Such a claim is not so much an assumption but rather more a logical consequence of the lack of any "causal" or other kind of

connection or relationship between the positions of the object at the different times. An object defined this way I call a *quantum object*; more particularly here, a Schroedinger object. Thus, in between measurements this kind of object exists in nonspace rather than in space.

2. The next question to be answered is: What happens if a classical object ($\ell = 0$) is at one of the locations (positions) of the random walk (quantum) object? The result is that the (quantum) object will be either in space—at the position of the classical object or in nonspace—at all of the other object positions. This procedure I call a *measurement*. This has nothing to do with interaction.

3. Since such an object has a position only when a measurement is made, it makes a spatial step only between measurements.

4. From this basis *exclusively* I then derive both the Schroedinger wave equation for a one-dimensional, noninteracting object, and also the Heisenberg Uncertainty Principle.

However, although the Schroedinger equation can be successfully derived on the above basis, the attempt to define mass in terms of equal step lengths turns out to be inconsistent with keeping the concept of mass within the concept of object. Therefore, in chapter VII, I again modify the concept of object, but this time from being a fixed step random walk, to one where the step lengths are the distances between the reversals of the random walk, thus giving a distribution of step length magnitudes, where now the mass is defined by the average of these step lengths. Amazingly enough, the Dirac equation results from this procedure. Such an object, still of course a quantum object, is called a Dirac object.

Basic to the developments in quantum mechanics developed here is the role of measurement. Thus, in chapter VIII the nature of nonspace and the role of measurement as integral to the concept of the quantum object, is defined and developed. Previously defined "objects" are thus seen to be approaches to the more fully formulated concept of the Dirac object.

In chapter IX the final statement is made on what, after all, is an object, although in chapters X and XI it is seen that even the concept of object, as I have said, has deeper ontological roots.

In sum, what has been done here is to start with the concept of a classical object, in which classical physics is expressed, an eminently practical, albeit incoherent view of the world, one in which the future, past, and present all coexist, or are at least codetermined, and to transform that view through a series of critical arguments into a possibly coherent idea of an object. Such a transformation shows both that determinism is a characteristic only of nonspace and imaginary time and that only a confrontation (e.g., a measurement) of an object with a classical object (itself a subset of the set of objects), resurrects time and space, and then only for an instant unless there is an interaction. Essentially, I show that the world as we experience it exists only because massive objects exist; but at each measurement "memory" disappears and an object's history begins again unless it becomes part of or is a classical object. Only because the objects around us are so massive contrasted to the basic objects of the universe, do nonspace and imaginary time not appear to us, who are also massive objects; only intervals of space and time are manifest, thus giving us the appearance of analyticity and the identity and continuity of objects in space and time. But the attempt to *define* the concept of object in a reasonable and coherent manner, which means to be able to determine its state at an instant of time for *all* time, at least between measurements, *and* its mass, may still be open. It will be shown that the determination of the mass of an object requires either a measurement of the locations of an infinity of identical masses in any interval, $\Delta t > 0$, or the location of the object after an infinite time. That is, there is, as is well-known, an inherent incompatibility between the space-time and momentum-energy (mass) properties of an object. Such an incompatibility may be overcome by interpreting the position probability density per unit time at a given time correctly (see discussion at the end of Chapter VII).

Perhaps one of the deepest insights achieved in this work is the realization that reality is "all possibilities" having, however, restrictions, and that it is *these* restrictions on nonspace that give rise to objects as well as to space and time.

It is the most faulty logic, even occasionally the crudest mysticism, something I abhor, that sometimes has led to fruitful results in physics. On the other hand, what is done here may comb out the logic of physics in giving us an ontology. I think such an attempt is necessary; the call of the world of physics to us is eventually for an ontology.

III

Classical Object

1. The concept of an object as the fundamental concept of physics is not normally discussed. What is discussed is rather the nature of mass, space, and time. The theme of this work is that eventually it is the concept of *object* which is the entry to the basis of physics and that the concepts of mass, space, and time are derivative from or secondary to the concept of object.

The concept of *object* is to be distinguished from that of (elementary) particles. Particles are of different kinds; e.g., protons, neutrons, electrons, neutrinos, pions, muons, etc.— even quarks. They all have properties such as mass, charge, spin, etc. But there is no statement as to *what* a particle is; i.e., what it is that gives rise to all these properties. The *it* is what I call the *object*. The concept of *object* is logically prior to the concept of particle; it is what *any* kind of particle is. Its nature is what is common to all particles.

2. In this chapter the concept of *object* is first developed as that of a *classical object*. The concept of a classical object derives from the Newtonian ideas that were the basis of mechanics from the time of Newton until the time of Einstein and are still conceptually viable today in much of physics, and persist in a lingering but particularly insidious way in quantum physics. Such ideas which define objects as small or "point" objects, which move or are at rest at a given position at a given time, are said to be "matter" or "substance," and possess "inertia," which is defined vaguely as "resistance to change." More specifically, an inertial mass relative to another inertial mass was defined by Newton's 2nd Law, $F = ma$, as holding only in inertial frames, or by Mach's definition of relative mass, $m_1 / m_2 = -a_2 / a_1$; but the concept of inertial mass itself remains quite vague in classical physics. In fact, the concept of

inertial mass plays its most significant role in early mechanics only where there is interaction. If there were no interactions, there would be much less of a role required for inertial mass in the definition of an object. But an object or "substance" would still exist. Thus, the concept of inertial mass cannot be identified with the concept of object in classical physics. But because *all* objects *do* interact, even if only gravitationally, all objects must have that property required by interaction, namely inertial mass. The further identification of gravitational with inertial mass, the foundation of The Theory of General Relativity, is still conceptually a puzzle. The only clue for the conceptual identity of these two concepts of mass is that since gravitation is the only universal interaction, its source, gravitational mass, must be identical with inertial mass, which is also universal for all objects. I shall not deal with this problem in this work, but I shall treat of the nature of inertial mass in later chapters.

3. We can call this early conception of an object from which the concept of the classical object is developed, the *Newtonian object*. The Newtonian object is said not only to possess the property of mass but also to be capable of existing in space and time. That is, a Newtonian object also has the properties of being at a position in space at an instant of time, space and time themselves having a (Euclidean) metric, thus allowing distances or intervals to be defined between space-time points. This is the basis of the ontology of classical physics; what exists are objects which are defined by these three properties, and a space-time plenum in which the objects exist.

4. I am careful not to say that space and time are properties of an object. This was certainly not Newton's view, although, as is well known, Leibniz did view space and time as *relations* between objects. In the Newtonian view, space and time, of course, always coexisted and need not have objects "in" them or have them "be present." Mass, however, being a property of an object, could not exist unless the object existed. Furthermore, objects do not exist "outside" of space and time (spacetime). Thus, the object can be said to be defined by its having a position in space and existing at an instant of time.

5. Eventually I shall claim that, even in classical physics, the concepts of space, time, and mass are derivative from the con-

cept of object. However, since the object was originally defined in terms of space, time, and mass, it is necessary to say something about these less basic concepts first. Actually, not too much can be said about them until we analyze the concept of object. What we can say immediately about space and time is the following:

6. By *space* is meant the set of space "points," each point being designated by 3 numbers, indicating 3 dimensions; by *time* is meant the set of time "points," each point being designated by a single number. The domains of the space-time numbers is the set of real numbers and thus extend from negative to positive infinity. Space and time have further properties, particularly that of metricity. That is, a distance function (Euclidean) is also defined between any two space-time points. This work deals only with one-dimensional space. Since, in Newtonian physics, space and time are taken as basic, primitive realities, other than rehearsing the Clarke/Leibniz correspondence, not much more can be said about them until the concept of object is further developed.

7. A Newtonian object, as stated previously, is said to occupy a point in space-time; that is, it has a position in space at a given instant of time. An object, by "occupying" such a space-time point does not "replace" it but is somehow there with it. This conceptual incoherence will be resolved in section 27.

Now, since an object occupies only a (single) position at an instant of time but may occupy the same position at many instants of time, the position, x, of an object is a *function* of the time, t, of an object; i.e., x = f(t). Every object in classical physics is uniquely defined and thus is distinguished from every other object through a function, x = f(t). (If the object were assumed to also have the property of mass at this point of the discussion, we would, for uniqueness, simply require all the masses to be the same.) Now, space and time are *independent* of each other; that is, there is no correlation or "connection" between any values of space and time. This means that there can be no restriction on any function relating space to time; that is, there is no function more special or preferred than any other. But this means that since *any* function can relate independent realities such as space and time, only a *special* or limited set of functions defines an object; another way of putting it, is that the presence or existence of an object is defined by a *restriction* on the set of all possible space-time functions.

If objects are non-interactive so that intertial mass may not be a necessary property of objects, they may then be defined *only* by a space-time function. If objects are not restricted to being only non-interactive objects, then they are required to have another property, say that of mass, in terms of which the interaction takes place.

8. Let us now turn to the question of the kind of functions by which objects are defined. An object is defined by a set of positions over the (time) domain of the object; i.e., the set of time instants over which the object exists. Two extreme cases exist; i. the domain of the defining function is infinite so that an infinity of positions is required to define the object; ii. the object's domain is a single instant, so that only the position at that instant is required to define it. Since this latter, of course, is not the usual way of defining an object, or has even been considered, as far as I know, let me consider the latter alternative first.

9. In this case, since the objects are all uniquely defined at a single time instant the positions of all the objects must be unique—the objects must all be at different locations. Since at a later instant of time these objects no longer exist but others may, these new ones also must have unique positions. Furthermore, there is no "connection" between objects at different times since these objects exist only at a single instance; there is no meaning to "the same object at different times" or "the identity of an object at different times." Furthermore there is no "disappearance" or "appearance" of objects, since, after a single instant that object no longer exists, there is nothing to disappear; similarly before the single instant there is nothing to appear. There is not even the *potential* to appear or disappear, since the object is only defined at an instant of time.

10. Suppose on the other hand that the (time) domain of the defining function is infinite; then an infinity of positions, one at each instant of time, would be required to define the object. Now, if the object exists at an infinity of instants, it must exist at each of these instants. Can it be said that it is

defined at an *instant* of time? If so, then at *each* instant of time the positions at *all* instants are defined even though there is no "connection" or "determinism" between the positions at different times. Therefore, since its positions are defined independently at *different* instants of time it cannot be defined at a *single* instant of time.

11. Therefore, the claim that the space-time function, x = f(t), of an object can be defined at a single instant of time, does not make sense or, in the former case, is totally inappropriate to any concept of a Newtonian object. Furthermore, in the latter case, none of the possibly many objects existing at t = t_0 at the same position, thus being indistinguishable one from the other, can therefore have unique functions. Since all these objects at *this* instant are identical, position being the *only* property by which they can be distinguished, there can be no *identity* between any one of these objects at another time with one of them at this time. Attempting to define an object in a finite time larger than zero, an intermediate case, simply partakes of the difficulties of both extremes. Thus, once again, one cannot even define an object at a given instant of time if its domain is greater than a single instant!

12. Such a conclusion certainly leaves us in a quandary. How can it be that the only ways to define an object in time, a simple, Newtonian object, fail, to say nothing about *predicting* the future position of an object, as is done in Newtonian physics even when forces are not involved? The resolution to this quandary is to be found in the restriction required on the defining function of the object. The restriction is informed by the following consideration:

All of classical physics requires motion to be smooth; to have discontinuous motions might even seem to be a definition of the miraculous; discontinuity seems to imply a disruption of "cause and effect." The laws of physics (or any laws, for that matter) seem to require smoothness and "connection." Perhaps here, in the requirement of continuity, lies the resolution of our quandary. Therefore, I impose the condition that no kinematical discontinuities of any kind can exist in classical physics; i.e., I will assume that the defining function of any object is "deeply continuous;" i.e., *analytic*. The classical intuition is that the notion of motion brooks no discontinuities. Thus, I chose, at first, the restriction of analyticity to define an object; this restriction defines what I call a *classical object*, and is in accord with Newton's and Leibniz's use of the calculus.

Now just because each object requires its own unique defining function does not imply the inverse; i.e., that a defining function implies the definition of an object. An object is specified by a function between values (numbers) of *space* and *time*.

13. Although the meaning of *analytic* is both well-known and widespread, I feel that it is required to further explicate it and the concepts upon which it is based in order to understand the concept of object identity in time and eventually that of the *classical object*. Although the concept of analyticity is an elementary one, it has significant consequences for classical physics that have hardly been exploited. Therefore, I wish to briefly review analyticity, emphasizing relevant aspects for the understanding of classical physics.

An analytic function is one whose derivatives are all continuous (and whose series expression converges). The concept of *continuity* is based on the concept of *limit function*. The concept of *limit function* is based on the concept of *limit* or *limit point of a set*. Thus, the basic concept upon which the idea of analyticity is eventually based is that of the limit point of a set. (The sets referred to herein are sets only of the rationals, irrationals, or reals. We shall first restrict ourselves to the set of reals.)

The *limit point*, t_0, of a set, $\{t\}$, is defined as a number such that *any* interval containing it contains at least one element of the set; therefore, it contains an infinity of other elements of the set. Thus, although t_0 may not be an element of the set it is not "isolated" from the set.

14. The *limit of a function at $t = t_0$*, $F(t_0)$, (the limit function) of a set, $\{f(t)\}$, is the limit point of the set, $\{f(t)\}$, at $t = t_0$. That is, the limit of a function at $t = t_0$ of a set, $\{f(t)\}$, is a number such that *any* interval containing it contains at least one, and therefore an infinity, of elements of the set in any interval containing $F(t_0)$. If the function equals the limit function at $t = t_0$ so that $f(t_0) = F(t_0)$ then the function is said to be *continuous at $t = t_0$*. However, it is possible that $f(t)$ may not

be defined there; i.e., the function may not exist at $t = t_0$, even though the limit function, $F(t_0)$, does.

15. If the limit, $F(t_0)$, of a function, $f(t)$, exists at $t = t_0$, then the function exists (is defined) in some interval, $\Delta t > 0$ containing t_0. Thus, implied in the statement that a *limit function* exists at a point, $t = t_0$, is that the *function* exists *both before and after that point* in a time interval, Δt. But even more than this is implied; it is also implied that the *function values* in $\Delta t > 0$ vary only slightly from each other. Thus, even if the function is *not* continuous at a given point, the existence of a *limit function* at a given point makes a considerable restrictive claim on the function both before and after that given time. Thus, to make the claim that the defining function of an object is analytic at a given time requires the position to be a continuous function of time *at* that time. But this means that the positions must not only exist before and after that time but are considerably restricted as to what they can be. In other words, the requirement of continuity *at* a given time restricts the possible values of position *after* that time.

16. Now, since analyticity puts a similar restriction on the derivatives of the function (that is, the position of the object as a function of time), it is required to analyze the concept of the derivative from which we define the *velocity* of the object.

In order to define velocity, $v(t)$, we first define another function, $g(t, t_0) = [f(t) - f(t_0)]/(t - t_0)$ (or in slightly different notation, $g(t_0 + \Delta t, t_0) = [f(t_0 + \Delta t) - f(t_0)]/\Delta t$). Such a function, of course, is not defined at $t = t_0$ since there $\Delta t = 0$ and the function is thus indeterminate. We now define the velocity function,

$$v(t_o) = v(t_o, 0) = \lim_{\Delta t \to 0} g(t_0 + \Delta t, t_0) = \lim_{\Delta t \to 0} \frac{f(t_0 + \Delta t) - f(t_0)}{\Delta t}$$

$$= \lim_{\Delta t \to 0} \frac{\Delta x}{\Delta t} = \frac{dx}{dt}$$

Such a function may or may not exist. If it does exist, then two things may be said:

a. The function, $f(t)$, is continuous at $t = t_0$, and b. The velocity is not $\Delta x/\Delta t$ but the *limit function* of $\Delta x/\Delta t$.

Furthermore, if v(t) is defined in some interval, $\Delta t > 0$, then the *velocity* at a given time is the same as the *average velocity* at that time; i.e.,

$$v(t_0) = \lim_{\Delta t \to 0} \left\{ \frac{\Delta x}{\Delta t} = \frac{1}{\Delta t} \int_0^{\Delta x} dx = \frac{1}{\Delta t} \int_{t_0}^{t_0 + \Delta t} v(t) dt = \bar{v}(t) \right\} = \bar{v}(t_0)$$

17. The existence of such a limit function means, of course, that for some $\Delta t > 0$, the ratio $\Delta x / \Delta t = [f(t_0 + \Delta t) - f(t_0)] / \Delta t$ is close to $v(t_0) = (dx / dt)_{t = t_0}$ This, then is an *added* restriction of the function at $t = t_0$; not only is the variation of the function restricted for some $\Delta t > 0$ but now so also the variation of the function *ratio*, $\Delta x / \Delta t$, is restricted as well. Now if a function is analytic in some domain, T, so that all its derivatives exist in the domain (and its Taylor expansion converges) then the derivatives at a *point* (in time) determine the function *everywhere in T*. That is, the restrictions of analyticity are so great that these restrictions defined *only* at a point uniquely determine the function everywhere else in the domain as long as the function is everywhere analytic there. Or, putting it conversely: in order to define a function analytic *at* a point, one must say everything there is to say about the function over its entire domain but state it at a single point. That is, the definition at a point of a function analytic in a domain T is equivalently the statement of the function at every point in the domain.

18. Instead of defining or specifying the derivatives at some point it is equivalent to define the function at a denumerable infinity of points in any interval, $\Delta t > 0$. (Thus the restrictions required by analyticity on a function are sufficient to reduce the number of its specifications from a non-denumerable to a denumerable set.) In either case, if the function is analytic, only a denumerable infinity of specifications, either at a point, t_0, or within an interval, $\Delta t > 0$, determine the function *everywhere* in T.

Thus, we conclude that an object which requires positions at many time instants in order to be defined, say over the interval $(-\infty, \infty)$, can be defined at a single instant if the defining function is analytic everywhere in the domain. The condition of analyticity means the restrictions on the possible positions are so severe that defining a denumerable infinity of positions on *any* interval $\Delta t > 0$ at some instant of time or, equivalently, defining all the derivatives at that instant is equivalent to defining *all* the positions and derivatives at *all* instants in the interval. That is, the restrictions are so severe that the future, as the past, is determined at the present.

19. Thus, if an object is defined by an analytic function in a domain, T, it is sufficient to define the function at a countable infinity of points in a region, $\Delta t > 0$. We then can say that the object defined in an interval $\Delta t > 0$ is the *same* or *identical* object existing elsewhere in T and distinguishable from all other objects. But all that is meant by the *same* or *identical* object is that its function values at all times are an analytic consequence of what they are in $\Delta t > 0$; an instant is the womb of all time if an object function is analytic.

20. Thus, what is called determinism turns out to be analyticity or identity and what is called *change* appears to be inexplicable in classical physics; i.e., change appears to be only the existence of time.

21. Zeno's arrow paradox, in the form that at an instant of time, the object (an arrow), since it is at a given place, is not moving, is now easily resolved. Implied in the statement that an object has a velocity at an *instant* of time, is that it exists not only in the present but in the past and future; furthermore, it is implied that not only the differences in its positions in some $\Delta t > 0$ are limited but so is the *ratio* of its positions to time differences in this $\Delta t > 0$; this is all implied in the statement that an object has a velocity at a given *instant* of time. That is, to say that an object is "moving" at an instant of time already is making a statement about the object's future position. *There is no "motion" in any other sense but of this analytic sort in classical physics*; the puzzle is more apparent than real.

22. Now suppose that the domain of the function (the time instants) is the set of reals (in some interval, finite or infinite) but that the range of the function (the positions) is only defined at either the rationals or irrationals. Then, the *limit function* (of the positions) may exist at each value of time, and the function might be continuous at the defined points, even though the *function* does not exist at the undefined points (re-

spectively, the irrationals or rationals) of position. That is, for a limit point of a function to exist requires the function to be defined at an infinity of time instants (points) within any $\Delta t > 0$ but not necessarily at *all* time instants in $\Delta t > 0$.

23. On the other hand suppose that the domain (time values) of the function is only the set of rationals (or irrationals). Then it is entirely possible that a function (the positions and their time derivatives) defined on this domain is *continuous everywhere* (in the domain), but defined over the domain of reals (*all* time instants) is *discontinuous everywhere*, so that the limit function exists nowhere. Such a function is: $f(t) = 0$, t rational; $f(t) = 1$, t irrational.

24. From these considerations we are led to the conclusion that the popular or usual idea of change as a "flowing," or that of an object not making sudden appearances or disappearances while being conserved, cannot withstand a critical analysis.(this of course was understood by Aristotle, Diorodus, et al.[2]) An object defined at an infinity of time instants or over a time interval as is done in this chapter, even if defined analytically, cannot produce the idea of change as a "smooth or continuous flowing."

25. From this discussion we then conclude the following:

a. The definition of an object as being a relation between values of space and time says that by *object* we mean, first of all, space and time exist; and secondly, that an object, being an object, is defined by a restricted space function of time. This function is restricted by being a space *analytic* function of time. Thus, there is, for all objects, a considerable restriction of space values on time values. Furthermore, this restriction is so considerable that if in any time interval, $\Delta t > 0$, the object s position is defined at a denumerable infinity of instants, or equivalently, by all its derivatives at a single instant, and if the function is defined as analytic in a domain T containing t_0, then the function is defined everywhere in T. Thus, a classical object is defined such that its derivatives at a *single* instant of time define it for all time. *Thus, there is no prediction, only definition; i.e., determinism is nothing but analyticity.* The second law of classical physics, F = ma, is the specification of the *interaction* of objects—a subject we do not deal with in this work-in terms of these kinematical variables. However, if F is an analytic function of position, then, once again, determinism is analyticity but now including interactions. Then, of course, the defining function includes the effect of the interaction. Therefore, the defining function of an object is analytic whether or not forces or interactions are involved. Thus, if all objects interact— as they do gravitationally—or when they do interact otherwise, the defining function of an object is also determined by the defining functions of other objects. Thus, if objects interact, they only can be defined in terms of each other.

b. The fact that an object is a space *analytic* function of time and thus not discontinuous in any of its derivatives allow us to determine, or so it seems, the future from a single moment in the present. The "predictability" is thus based on the idea that the motion or kinematics of objects are "deeply smooth and continuous"; i.e., analytic.

c. If there are no forces, then instead of an infinity of derivatives it appears that only two are required to define an object. But the statement that there are no forces on the object is that since F = ma, a = 0 for all time; thus, all the derivatives higher than the first are already given as zero.

d. Does one somehow *find* an object already defined at a given instant or does one *create*, fix, establish, determine the values of the position and derivatives at a given time instant; i.e., does one discover or establish the initial conditions of an object? If it is the latter then it would appear that the object at times previous to the initial time is not defined. But, according to our presumptions *all* objects (in classical physics) are defined—otherwise there is no meaning to the concept of object. Therefore, all that is done is to *find* or *discover* a given object at the time of "establishing" the initial conditions. It is simply, in classical physics, that the "establishing" or "creating" of the initial conditions is *itself* determined. We will see that in quantum mechanics this is not so.

26. How is time distinguished from other variables, such as space? Points in time, just like points in space (on a line) are ordered; thus the characteristic of order is not sufficient to distinguish time from other variables such as space (one-dimensional). A distinction between space and time is made through the concept of object, an object being defined as a position *function* of time, so that time is the *ordering* of position values. However, this by itself gives no way of distinguishing past from future, either for a single object or collection of objects; that is, classical physics, as described here, cannot produce a basis for defining time. Since the concept of time is basic, although not sufficient, as shown above, to define the concept of change, classical physics then can produce no basis for the concept of change. Such a basis can only be produced, if at all, by another physics, such as quantum mechanics. The concept of change, itself, will then probably be considerably different from how we normally understand it. It is not that I deny that change exists, but simply that I do not find it, as yet, a coherent concept—certainly not in classical physics. I discuss the nature of change briefly in section 96.

27. In the work done so far space and time have been assumed to be external to, outside of the object, existing whether or not the object exists, normally independent realities related to each other through the object itself being defined as a space function of time. I have, in section 7, claimed that the concept of external space and time is incoherent. I now give my reasons for this claim.

What does it mean to say that an object "occupies" a certain *location* or is at a given *point* in space? It must mean that it is "there," at that location. That is, it must mean that *its* location is identical with the spatial point, that there is a coincidence. That is, it must mean there is a coincidence between *its* location and that of the spatial location or position. This means then that the object itself has the *property* of spatial location or position (and similarly of time instants) distinct from external space (and time). But then the apparent need for external space and time no longer exists; all space and time values are properties of objects. But not only is there no need, but the very concept of space and time being realities external to objects is not derivable from any measurement procedure since all measurements are done on objects and its properties. It seems that space and time *outside* of objects, just like the long gone ether, are hangovers from a Newtonian legacy.

28. The other basic concept of classical physics besides the analyticity of the object function is the concept of frames of reference. That is, space and time points have order but they do not have values; the *value* of a space or time point is defined only if there is an additive identity element in the set of points. That is, the values of space points exist only relative to the (spatial) points or locations of other objects; the value of a spatial point or location of an object is only defined relative to or with respect to the spatial point or location of some other object. Thus, the *value* of the point or location of an object is the signed (positive or negative) *distance* to some other object. (Zero distance if the other object is the same object.)

This is true not only for position (and time) but also true for the time derivatives of space, such as velocity, acceleration, etc. That is, an object has *values* of these properties only with respect to those properties of some other object. Thus, $v = dx/dt$ is not yet defined as the velocity of an object until it is specified with respect to what other object it is being taken. Thus, to define an object, even if there are no forces or interactions, requires generally an infinity of other objects to act as frames of reference.

29. The most coherent view of the classical world then is that its basic entity is a *classical object*, which is defined as having three properties; mass (not yet discussed), spatial position, and time instant. An object, specifically a *classical object*, is then seen to be the *only* kind of entity of the (non-interactive) world. Space and time (and also mass, as will be shown in the next chapter) are not separate entities but rather an operational statement or definition of what an object is.

We see that the original quandary of how to define an object is resolved by requiring the defining function of the object to be analytic; then we can uniquely define an object at a given instant in time for all time. However, the implication of this is that if there are to be no discontinuities and thus no "miracles" in the universe, then all that *will* exist (and *has* existed) can be defined at an instant; there seems to be no in between. Newton's idea that space and time exist independently of objects turns out to be both unnecessary and incoherent.

In the next chapter, on relativity, I will develop the concept of object from that of classical object into that of a relativistic object in order to overcome not yet mentioned difficulties presented by the concept of classical object, both conceptual and phenomenal. I will then continue this conceptual development into the later chapters, thus establishing the concept of *object* as the basic ontological reality, at least as far as a single object in non-interactive physics is concerned.

IV

Relativistic Object

30. The original derivation of the special theory of relativity by Einstein is based on a requirement derived from the theory of electromagnetism, that the velocity of light, assuming light to be a participant in the universe, is the same in all galilean frames of reference. This is a necessary condition for the theory and consequently is a basis for the Lorentz transformation; yet, one might wonder what a particular phenomena such as light has to do with a space-time transformation. That is, if the Lorentz space-time transformation holds, then the velocity of light must be the same in *all* galilean frames of reference. And certainly if the velocity of light is the same in all frames of reference, then the space-time Lorentz transformation must hold. But the Lorentz transformation will hold whether or not light even exists. The source of the Lorentz transformation can arise only out of the nature of the object itself or the nature of space and time and not out of phenomena external or irrelevant to it. This is implicitly recognized by the later formulation of the special theory in its substitution for the electromagnetic postulate, the postulate that the maximum speed of all objects is finite (and that the space-time transformation between any two galilean frames must be unique).[4] However, there never has been an explanation of *why* there is necessarily an upper bound to the speeds of all objects, thus lending an air of incompleteness to the theory of relativity, at least to some of us. However, an explanation does exist, and it derives from a rather unexpected source—from the attempt to define an object in classical physics.

31. Although in the previous chapter I have resolved the quandary of defining an object uniquely at a given instant of

time by requiring its defining function to be analytic, there is an equally daunting second quandary in classical physics I have not yet mentioned. Even if the defining function is analytic, there is no upper bound on any of the derivatives, particularly the speed. Since there is no reason why one speed or velocity is preferable to any other, the probability distribution of all possible velocities of the objects when chosen at random is independent of the velocity, and thus is "flat." Thus, in any time interval, $\Delta t > 0$, except for a very rare few objects, since the range of possible velocities is infinite, the object must be at a distance, x, from its original position larger than *any* given distance; that is, for any given distance, D, x > D (by very few is meant a zero percentage of all possible velocities). Thus, except in extremely rare cases, it is not even possible to define an object. In order to resolve this quandary it is necessary to restrict the defining function of an object so that the displacements made by a given object in a given time is limited; *that is, there must be a maximum speed to all objects.* It is rather surprising that the basic rationale for the special theory of relativity, the existence of a maximum speed for all objects, arises from the need to define the concept of object.

Thus, in order to define the concept of object, it is required to define it so that it has a maximum speed. However, just like a classical object, it must also be definable at an instant of time; therefore, it must also be analytic—no discontinuities in its function are to be allowed. The problem for solution in this chapter is to determine what the nature of an object is so that its defining function can still be in some way "deeply continuous" and yet have a maximum speed. Such an object will be called a *relativistic object.*

32. The guide to doing this is to redefine the classical object not by a continuous function, but rather by a step function random walk. It is in this manner that it will be possible to introduce a maximum speed. Only then, by allowing the "step length" to approach zero, will we be able to get back to the required analytic function. This procedure allows us to distinguish between (instantaneous) velocity and (instantaneous) average velocity; it will be the average velocity that will have a maximum magnitude.

The defining function of an object is now written as x = g(t), where t = nτ and x = sℓ ; ℓ is the "distance" moved in one step and τ is the "time" it takes; n is the number of steps taken by the object, an integer, and s is the net number of steps from the starting position, an integer. Thus, if n changes by one unit, s can change by +1 or –1 units. It is in this manner that the difference in locations (the distance "moved" in any given time interval) is limited, and we resolve the paradox mentioned in section 31.

We see that the defining function for a relativistic object is nothing but a binary or simple random walk function. But just as in the case of a classical object, here too, because we have a highly select subset of all possible functions on the number set,{s,n}, we can say x designates a spatial point and t designates a time instant since an object is defined by a *spatial* function of time. Then $t_{n+1} = t_n + \tau$, so that $\tau = t_{n+1} - t_n$ is the time interval between consecutive (or adjacent) time instants and $x_{n+1} = x_n + \ell$, where ℓ is the distance between adjacent spatial points. Thus, $\Delta t = n\tau$, n a positive integer, and $\Delta x = s\ell$, $|s| \leq n$. Thus, the change in distance per unit time is bounded,

$$\frac{|\Delta x|}{\Delta t} = \frac{|s\ell|}{n\tau} \leq \frac{\ell}{\tau} = c, \quad c \text{ constant.}$$

A given object is still defined by a given function but now the function is no longer an analytic function but a sequence of "steps," x, where,

$$x_n = g_n(t) = \sum_{i=1}^{n} (\Delta x)_i = \ell \sum_{i=1}^{n} f_i = s\ell, \quad \text{where } f_i = \pm 1 \text{ and } |s| \leq n.$$

That is, an object is now defined, as stated above, as a binary random space walk of steps all of equal magnitude on equal time intervals.

33. Now, since x = g(t) is a *distance*, then, just as is required for classical physics, x must be defined relative to an object, a spatial frame of reference. For classical physics, if x = f(0) = 0, then, for any Δt = 0, x is the difference in location or position from x = 0 and it can have any value; here, for Δt = τ, x = +ℓ or –ℓ only; i.e., there is one of only *two* possible values of x relative to another object. And since the steps can only be ±ℓ , then the

velocity relative to any other object can only be one of the two values, $\pm c$.

34. For a random walk object where the function is

$$x = \ell \sum_{i=1}^{\infty} f_i, \qquad f_i = \pm 1,$$

the average velocity is

$$\bar{v} = \lim_{n \to \infty} \left[v_n = \frac{x_n}{t} = \frac{\ell \sum_{i=1}^{n} f_i}{\tau n} = \frac{c \sum_{i=1}^{n} f_i}{n} \right] \leq c \text{ and } \geq -c$$

That is, the "average" velocity for a random walk, $\ell > 0$, as stated above, can only be defined over an infinite number of steps since the time domain is infinite; higher "derivatives," such as acceleration, etc., cannot exist since it takes infinite time to define velocity. Now if v_n has a limit, so that $\bar{v}$ exists, we see that $\bar{v} = cm$, $|m| \leq 1$. Thus for a random walk, the average velocity is bounded. But as we can see, since $\tau > 0$, an infinite time is still required to define an object and its average velocity. Thus, since the average velocity, $\bar{v}$, is defined in terms of the velocities, $\pm c$, $\bar{v}$ is also defined relative to other objects.

35. We wish now to define the object so that, just as with a classical object, its average velocity is defined at an *instant* of time; that is, it is required that $\tau \to 0$. Assuming that $\lim_{\tau \to 0} c = c$, then since $t = n\tau \to 0$, τ must approach zero faster than n approaches infinity. That is, take a random walk so that in the averaging time, t ($= \Delta t$), there are many steps. Then as Δt decreases, the size of the steps is allowed to decrease even more rapidly so that even though t is approaching zero, the number of steps in the time interval is approaching infinity; e.g., set $\tau \propto 1 / n^2$. Then the (instantaneous) average velocity can be defined at an instant of time

$$\bar{v} = \left[\lim_{\tau \to 0} c \right] \left[\lim_{n \to \infty} \frac{\sum_{i=1}^{n} f_i}{n} \right] = cm,$$

where +c or –c is the (instantaneous) velocity at that same instant of time. It is in this way that both velocity and average velocity can be defined at each instant of time; the former being highly discontinuous, having only values ±c, and the latter being continuous, and now even the higher derivatives can be, once again, definable and analytic in time.

As we have seen, a particular object is defined by the binary ordering of the signs (+ or –) of the step length or interval, ℓ, as it approaches zero. That is, wherever there is this randomizing of such an interval there is an object; the randomizing *is* space and time. Therefore, wherever there is an object there is *both* space and time and *only* there. (In the next chapter I will show the consequences of keeping ℓ (and τ) larger than zero.)

36. In the preceding sections the random walk was defined in terms of a distance step of length ℓ taken in a "time" τ. I then claimed that c = ℓ/τ was a constant for *all* values of τ. Such a claim may not appear so arbitrary if it is acknowledged that ℓ and τ must actually be the same or identical entity since no distinction has been made between them. What is distinct is not ℓ and τ, but $\ell' = \pm\ell$ and τ. Thus, while τ is always positive, ℓ' is either positive or negative. τ *then* may be taken as a step in time, ℓ' as a step in space if sufficiently many steps are taken so that τ and ℓ' can be distinguished. If steps are taken all of which are +ℓ or all of which are –ℓ, then space and time are not distinguishable in this theory. However, this would be a highly unusual situation; i.e., ℓ' and τ steps are distinguishable (n sufficiently large) and thus space and time "arise" from "pre-space-time" objects as distinguishable realities. We can then, since they are distinguishable, chose different units for them and make c units of ℓ equal to one unit of τ so that ℓ = cτ. Thus c has no physical content, but c' = ±c does. This approach is discussed further in the chapter on measurement.

Now the presumptions made in this section, if correct, expose the source of the special theory of relativity as being *the identity of the magnitudes of the space and time steps, no matter how small.* The constancy of the velocity of light in all galilean frames or the fact that the velocity of all objects is bounded is not the basis of the special theory of relativity, but rather a consequence of it. The special theory of relativity expresses itself in a particular space-time transformation—the Lorentz transformation between galilean frames of reference.

37. Now, from these results, how do we get the special theory of relativity, particularly, as expressed in the Lorentz transformation? The space-time transformation between different galilean frames in the special theory of relativity (the Lorentz transformation), is based on two ideas:

a. The transformations of space-time coordinates between any two galilean frames is independent of the galilean frame and of any particular space or time values. In the usual or standard derivations of the theory, since v is continuous, $v(t) = \bar{v}(t)$, so it makes no difference whether the galilean frames are defined in terms of the velocity or the average velocity, as shown in section 16. However, in this theory, where the object "moves" so that $v(t) = \pm c$ only, the galilean frames must be defined in terms of the average velocity, $\bar{v}(t)$; how this is to be done will be shown in section 39. Therefore (a) will be true of this work, as it is in the standard theory as long as $x = \bar{v}(t)t$ is analytic. Because of the equivalence of reference frames expressed in (a), the most general form of coordinate transformation from one galilean frame to another is, as is well known, $x' = g (x - vt)$. $t' = g [t - (v / c^2) x]$, where $g = [1 - (v^2 / c^2)]^{-\frac{1}{2}}$ and where c is either finite (real) or infinite (v is now designated as the average velocity instead of "$\bar{v}$"). c, as can be seen from the transformation equations, is the maximum object speed. If c is infinite, we have the galilean transformation; otherwise we have the Lorentz transformation.

b. There is a maximum speed for all objects independent of any galilean reference frame. *But this is exactly what I have shown.* We see then that since the (instantaneous) velocity is always $\pm c$, and that the (instantaneous) average speed is always bounded, $|v| \leq c$, that we immediately get the Lorentz transformation. That is, by giving a *justification* for the existence of a maximum speed to all objects *I have derived the special theory of relativity*–more specifically, the space-time Lorentz transformation. No recourse to an irrelevant phenomena, such as the velocity of light, is required. (In fact, the characteristics of electromagnetic waves, instead of being a *basis* for relativity are *required* to satisfy the theory.)

38. From the theory developed here we not only can derive the Lorentz transformation, but because we now understand the source of the maximum speed of all objects, we can get insights into the underlying physics of the Lorentz transformation not available to us before. This physics arises because an object has *both* a velocity and an average velocity with respect to every other object and thus with respect to every frame of reference. In order to specifically exhibit these insights, consider $\ell \to 0$ but not yet zero. Then in the step time, τ, each object—and therefore, frame of reference—has not only an average velocity (approximately) with respect to every other object but also a velocity, $\pm c$, with respect to every object. The problems to solve now are: how do these velocities transform with respect to different galilean frames and from this, *without using the Lorentz transformation*, how do the average velocities transform? This should result in, of course, the well-known transformation formula,

$$v_3 = \frac{v_1 + v_2}{1 + \dfrac{v_1 v_2}{c^2}}$$

which is derived from the Lorentz transformation.

The velocity of an object with respect to any other object can only be +c or –c. We now determine how these velocities transform relative to different galilean frames and thus get further insights into the physics of the Lorentz transformation, particularly time "dilation." Furthermore, since the average velocity of an object is determined by the difference of the number of positive and negative values (steps) in its defining sequence (function) then, to be able to determine a formula for the transformation of the *average* velocities from one frame to another—that is, the addition of velocities formula for relativity—it is necessary to determine how the velocities—or steps—

themselves transform. It should be noted that in order to define distance and velocity (section 39) the step length, even when it is vanishingly small, must have exactly the same magnitude with respect to all frames of reference, so that any time "dilation" or length "contraction" in going from one frame to another is a result only of different numbers of "events," or "steps" for the same object referred to different frames of reference. This is because, as stated above, the relative velocity of objects is *defined* in terms of a fixed step length magnitude. Thus, it is only the *number* of events relative to a given frame of reference we need concern ourselves with.

39. Suppose object A has a velocity +c or –c relative to object B and object B has a velocity +c or –c relative to object C. Then what is the velocity of A relative to C? Since it can only be ±c, it is obvious that the transformation cannot be galilean; the values of the velocity do not form a group with respect to addition. We will say that in this case the velocity of A relative to C is respectively +c or –c. On the other hand, suppose that A has a velocity of +c or –c relative to B but B has a velocity +c or –c relative to C. Now, what is the velocity of A relative to C? In this case there is no basis for choosing either +c or –c. We are then left with the only remaining alternative which is that no *event* occurred, no step took place. An example of a sequence of such transformations is:

Sequence of steps:	1 2 3 4 5 6 7 8
Velocity of A relative to B:	+ + + − − + + +
Velocity of B relative to C:	+ − − − + − + +
Velocity of A relative to C:	+ − + +

We see that between steps (1) and (4) there are 3 steps in row one, but only 1 step in row three. Since the step-length (and step-time) are presumed the same between all events and in all frames of reference we conclude that a clock measuring these events in frame B would measure three times the time that a clock in frame C would measure. Furthermore, a clock would measure equal times between any two events in any given frame, particularly in row three. Thus, we see that the non-group nature of the binary values of the velocities leads to time (and space) intervals themselves being frame dependent.

40. I now derive the transformation between velocity frames for the (instantaneous) *average* velocity; it is the *average* velocities in the theory developed here that are the velocities in the standard theory of relativity.

Consider now a given (relativistic) object defined by a sequence of steps having a given v and a vanishingly small ℓ. Two other terms, p = [1 + (v/c)] / 2, q = [1 − (v/c)] / 2 are defined so that p + q = 1; p is the fraction of positive steps in the sequence and q is the fraction of negative steps. By solving for v, we get v = (p − q)c, where, as can be seen, c is not only the speed (at each step) but also the maximum average speed.

41. In section 39, I defined the transformation of binary valued velocities from one velocity frame to another. In this section I will show, based on this understanding only and not using the Lorentz space-time transformation, how the formula for the transformation of *average* velocities (the sum of velocities formula) may be obtained. I exhibit this relationship explicitly in a derivation of the sum of velocities formula,

$$v_3 = \frac{v_1 + v_2}{1 + \dfrac{v_1 v_2}{c^2}}$$

Suppose now that we are given three objects such that the velocity of object B, v_1, is defined relative to the velocity of object A; the velocity of object C, v_2, is defined relative to the velocity of object B; and the velocity of C relative to A is v_3. Then, what is the relationship among v_1, v_2, and v_3? That is, given v_2, the velocity of a random walk relative to frame B, what is it relative to frame A if the relative velocity of the frames (as a random walk) is given?

I now define the steps in v_3 given the steps in v_1 and v_2. If, at a given time, the steps of both v_1 and v_2 are +, then the step in v_3 will be +; if the steps in v_1 and v_2 are both −, then the step in v_3 will be −. If on the other hand, the steps in v_1 and v_2 are of opposite signs, then *no* step will occur in v_3; that is, even though a step occurs in v_1 and v_2 *none* will occur in v_3. Whether or not a step occurs in a frame is not *absolute* but depends on the frame as indicated. To repeat, it is seen then that here is at least *one* source for the difference of time intervals in different frames—the number of steps *even though τ is the same in all frames.* Now, since the only steps in v_3 come from v_1 and v_2, both being + or −, and the steps are independent of each other, the probability, p_3, of the step being + is $p_3 = p_1p_2 / (p_1p_2 + q_1q_2)$, where p_1p_2 is the probability of a + step in both v_1 and v_2 and $p_1p_2 + q_1q_2$ is the probability that for steps in v_1 and v_2 there will be a step either ± in v_3. (The probability of p_1q_2 or p_2q_1 contributing to a step in v_3, as stated above, is zero.) Then, substitution of the expression for the p's and q's in terms of the v's, gives us

$$v_3 = \frac{v_1 + v_2}{1 + \dfrac{v_1 v_2}{c^2}}$$

the relativistic formula for the addition of velocities. Such a formula, as we can see, is independent of the value of ℓ.

This derivation requires the step lengths of the objects to be of the same magnitude. But suppose that the objects had different step lengths. Would the same formula for the addition of velocities still be true? Suppose that, even if the objects had steps of the same magnitude, these steps did not begin (and end) at the same time? Will the formula for v still be the same? The answer is yes. One simply considers that interval which is the largest common divisor of both step lengths as the (new) step length of both objects (assuming commensuratility). Since the v's are independent of the size of the step length, the same formula immediately follows.

Now, since $τ \to 0$, so that the average velocities are (approximately) defined, then it is possible to define x = g(t) as an analytic function, thus resulting in the special theory of relativity; the physical source of the above formula, namely, that the number of steps is frame-dependent even though the values of ℓ in all frames is the same, still holds.

It is seen that since there are fewer events (steps) in adding velocities randomly in the manner above then in the velocities being added as they would be done classically, that the distances (steps) being added are less than the sum of the distances the objects have moved whose velocities are being added, thus giving an (average) velocity less than we get in non-relativistic physics.

Furthermore, as was stated earlier, the time "dilation" or the "slowing down" of moving clocks can also be understood on the same basis. A moving clock has fewer events or steps than the stationary clock in the latter's time, thus producing a time "dilation."

42. The discovery that an object can have two different kinds of instantaneous velocities, one having values ±c and highly discontinuous in time, and the other, the average, v(t), which can be analytic in time, is, as shown above, the basis of the special theory of relativity. At the same time, the actual definition of the object as a highly discontinuous sequence of ±c values for the (instantaneous) velocity is not uniquely specified for a given analytic function defining the object. That is, although the function specifying the object, x = g(t), may be analytic, there are an infinity of ±c object specifications or paths that satisfy any given analytic function, x = g(t). Thus, since relativity arises in the above manner, analytic specifications of an object are only an *averaging* of one possible +c specification or path. We thus lose some of the completeness and specificity of the concept of the classical object.

It should be noted that the above basis for relativity theory is also an explanation for the eigenvalues of the velocity operator in relativistic quantum mechanics being only ±c.

43. I now derive the conservation of the momentum-energy 4-vector. The purpose of this derivation, which is different from the usual derivation, is to show how the concept of mass is required for relativity theory.

The Lorentz transformation of space-time points between galilean frames can be equivalently expressed as a rotation in a Minkowski (pseudo-Euclidean) space. Therefore, I can write,

$$\sqrt{-1}\ \mathbf{u} = \mathbf{e}_1 x + \mathbf{e}_2 y + \mathbf{e}_3 z + \mathbf{e}_4 \sqrt{-1}\ ct = \mathbf{r} + \mathbf{e}_4 \sqrt{-1}\ ct$$

$$= (x,\ y,\ z,\ \sqrt{-1}\ ct),$$

where $\mathbf{u}$ is the space-time 4-vector, $\mathbf{r}$ the space vector and $\mathbf{e}_1$, $\mathbf{e}_2$, $\mathbf{e}_3$, $\mathbf{e}_4$ are the space-time unit vectors; the domain of (x, y, z, t) is $(-\infty, \infty)$.

If we take the difference of 2 space-time points, we get, where $\Delta\mathbf{r} = \mathbf{r}_2 - \mathbf{r}_1$, $\Delta t = t_2 - t_1$,

$$\sqrt{-1}\ \Delta\mathbf{u} = \Delta\mathbf{r} + \mathbf{e}_4\sqrt{-1}\ c\Delta t = (\Delta x,\ \Delta y,\ \Delta z,\ \sqrt{-1}\ ct),$$

where the domain of $(\Delta x,\ \Delta y,\ \Delta z)$ is $(-\infty,\ \infty)$ and that of Δt is $(0,\ \infty)$.

If we divide by $\Delta u = |\Delta\mathbf{u}| = \gamma(\Delta t)c$, where $\gamma = [1 - (v^2/c^2)]^{-\frac{1}{2}}$ then,

$$\sqrt{-1}\ \frac{\Delta\mathbf{u}}{\Delta u} = \sqrt{-1}\mu = \gamma\ \frac{\mathbf{v}}{c} + \mathbf{e}_4\sqrt{-1}\gamma = \gamma'\ \frac{\mathbf{v'}}{c} + \mathbf{e}_4\sqrt{-1}\gamma'$$

where μ is a unit vector and $(\gamma,\ \mathbf{v})$ and $(\gamma',\ \mathbf{v'})$ refer to any two galilean frames of reference.

44. I now show that if there can exist a system of two or more interacting objects, that a concept such as mass, m, is required.

Suppose that we have a system μ_0 of *two* objects, μ_1, and μ_2, where $\mathbf{v}_1 = -\mathbf{v}_2$. Then, if m is not introduced, we have,

$$\sqrt{-1}\mu_0 = \sqrt{-1}\ [\mu_1 + \mu_2] = 2\sqrt{-1}\gamma\mathbf{e}_4$$

and thus $|\mu_0| = 2\gamma$. Since, for a given system, μ_0 is a constant and γ varies with the interaction, there is a contradiction, and another parameter such as mass, m, must be introduced. If m is introduced then we see that although the mass of the system, m_0, is a constant, the mass of each object, m, is a variable, as required. That is,

$$m_0\mu_0 c^2\ \sqrt{-1} = m_0 c^2\sqrt{-1} = 2mc^2\gamma\sqrt{-1}\mathbf{e}_4 \Rightarrow m_0 = 2m\gamma$$

so that, since m_0 is constant, and $\mathbf{v}$ is variable, m must also be variable. Thus, m *must* be a parameter of the object. I have thus given a rationale for the existence of mass, namely, that if interactions can exist, mass must exist. Generally, we can write,

$$m_0 c^2\sqrt{-1} = \gamma m\mathbf{v}c + \mathbf{e}_4\gamma mc^2\sqrt{-1} = \gamma'm\mathbf{v'}c + \mathbf{e}_4\gamma'mc^2\sqrt{-1}$$

Or, in other notation,

$$\sqrt{-1}m_0 c^2 = c\mathbf{p} + \mathbf{e}_4\sqrt{-1}\ E$$

where, $\mathbf{m}_0 = \mu m_0$, $\mathbf{p} = \gamma m \mathbf{v}$, and $E = \gamma mc^2$.

This, of course, is the Lorentz invariant momentum-energy equation. But it is also a *conservation* equation; that is, for an isolated system m is fixed. Such a *conservation* law only arises because m is fixed; i.e., it is not only Lorentz invariant but also an invariant of the object, fixed in time.

45. In conclusion it is seen then that relativistic objects derive their fundamental nature from the fact that the magnitudes of step length and step time are the same; that is, they are a *single* reality, and the only way in which step length and step time differ is by the fact that the former is random relative to the latter. That is, the fundamental uniqueness in Newtonian physics of two physical realities, space (length), and time is reduced, at least in magnitude, to only one. In the next chapter, on the random walk object, I will identify mass with the step length or step time; thus, even *before* we get to quantum mechanics, the three basic physical realities of physics, mass, space, and time, are found to be based on a *single* physical reality, that of the pre-space-time "step." Furthermore, such a "step," the basic physical reality, which does *not* depend for its value on any frame of reference, appears to imply the necessity of a universal interaction, such as gravitation. Could this be the source of the identity of inertial and gravitational mass?

V

Classical Random Walk Object

46. In the previous chapter, on relativity, it was found that in order to define an object it was required that there be a maximum speed to all objects. This requirement, which gives rise to the theory of relativity, does so through the fact that, in the "small," space and time are identical—or rather that space and time both arise from the same source, a kind of pre-space-time. These pre-space-time intervals, when ordered, are time intervals and when randomized as positive and negative intervals are space intervals. In order to make the function defining the object analytic it was then required to allow the size of these steps to approach zero. Because the "velocity" of these steps is now everywhere or almost everywhere discontinuous, being only ±c, *the (instantaneous) velocity and (instantaneous) average velocity are no longer equal.* Therefore, to repeat, when I say that the object's defining function is analytic, I am referring now only to the object's *average* velocity and its derivatives; the velocity itself is always either +c or −c but which it is, is not defined. There are an infinity of such defining functions for any given average velocity.

It will be seen in the next chapter, on quantum mechanics, that the relativistic object itself, and thus classical physics, is still incoherent and that the concept of object requires further analysis and development. However, we must prepare for this task by first doing some work in the application of a random walk to the concept of object, where the step length does not, as in the previous chapter, approach zero. This is because, as stated above, the step length is the extra parameter required to identify with mass in order to integrate the latter into the concept of object. Therefore, I will consider a random walk not only having a maximum velocity but also one having a definite step length, $\ell > 0$. Surprisingly, we will find that many results—e.g., the de Broglie relation, a kind of Uncertainty Principle—derive, not from quantum physics, but from a new conception of the object.

47. In defining the relativistic object I started out with a random function, $v_n = \ell \sum_{i=1}^{n} f_i / n\tau$, and then let $\ell, \tau \to 0$. In order to further exhibit the nature of an object—both relativistic and quantum—I will consider first a random walk defined not by (x,t) but one whose sample space is the set of integers, (s,n), $s \leq n$, where n, being the number of steps, is positive, and s, being the sum of the random steps, is either positive or negative. The relationship between the points of the two sample spaces is $x = s\ell$ and $t = n\tau$. I will then consider the random walk whose sample points are (x,t). The purpose of considering a random walk over two such distinct sets of sample points is to be able to evaluate the effect of introducing the step length ℓ—which can be of any magnitude—on the concept of object. In any case, as long as $\ell > 0$, so that analyticity is lost, so is the ability to define an object at an instant of time and thus our classical ontology is lost. For this reason, the work in this chapter can at best be only a preparation for the chapter on quantum mechanics. Nevertheless, it is a necessary preparation.

48. The probability distribution of the sum of n steps each having a probability of p to be +1 and a probability of q to be −1 is

$$P(s) = \frac{n! \, p^{((n+s)/2)} \, q^{((n-s)/2)}}{\left(\frac{n+s}{2}\right)! \, \left(\frac{n-s}{2}\right)!}$$

For equal preferences, that is, where $p = q = \frac{1}{2}$, we then have,

$$P(s) = \frac{n!}{2^n \left(\frac{n+s}{2}\right)! \, \left(\frac{n-s}{2}\right)!}$$

For n,s → ∞ (such that $s^3/n^2 \to 0$), P(s) can be approximated by the normal distribution,[5]

$$P(s) = \frac{e^{(-s^2/2\alpha^2)}}{\sqrt{2\pi}\ \alpha}, \qquad \alpha^2 = 4npq$$

or, for $p = q = \frac{1}{2}$,

$$P(s) = \frac{e^{(-s^2/2n)}}{\sqrt{2\pi}\ \sqrt{n}}$$

The standard deviation for such a distribution is $\alpha = \Delta s = \sqrt{n}$. The standard deviation of Δs per unit n is $\Delta w = \Delta s/n = 1/\sqrt{n}$, so that

$$(\Delta s)\,(\Delta w) = 1$$

Now, by re-introducing the "physical intervals" ℓ and τ, we change back to the original sample space (x,t) and get for the random walk,

$$P(x) = \frac{e^{(-x^2/2(\Delta x)^2)}}{\sqrt{2\pi}\ (\Delta x)}, \qquad (\Delta x)^2 = n\ell^2$$

where this normal distribution can be an excellent approximation to the binomial distribution (in x and t) even though x and Δx may be relatively small; this is because s and n can still be quite large. If now ℓ and $\tau \to 0$, once again, there might be analyticity and we might have a relativistic object. But if ℓ and $\tau > 0$, then there is no longer analyticity and therefore we can *no longer define an object*. Thus it can no longer be claimed that an object is the *same* object at two different instants of time, unless there is *either* only a single object *or* if there are two or more objects present, they must be at least a distance x = ct from each other. The only connection now between an object at $(x = x_0, t = t_0)$ and the "same" object at $t = t_0 + \tau$ is that at the latter time it is at one of the positions $x = x_0 \pm \ell$

That is, since we still have an object, there must be the above restriction on possible space-time functions. But it can no longer be claimed that the restriction still produces objects having defined functions; identity no longer exists for an object so concerned. Identity, after all, is simply a matter of the definition of an object; in classical physics, as we have shown, identity can be specified for all of time at an instant; by contrast, for a pre-quantum (random walk) object, it requires *all of time*. There is no *property* of identity, no quality of *sameness* from instant to instant. At best we can say that the number of objects is invariant with respect to time—and *that* only for non-interacting objects—and that the mass of a system is also invariant with respect to time.

49. I can now write dispersion relations on the (x,t) sample space and the uncertainty principle for x and v,

$$(\Delta x)^2 = \ell^2 (\Delta s)^2 = n\ell^2 = \ell\, ct$$

$$(\Delta v)^2 = v^2 = \frac{x^2}{t^2} = \frac{(\Delta x)^2}{t^2} = \frac{\ell c}{t}$$

Then,

$$(\Delta x)(\Delta v) = \ell c$$

Thus, $(\Delta x)(\Delta v)$ is, as is $(\Delta s)(\Delta w)$, a constant for all n (and t). This, of course, is simply a result of a random walk.

The step length and the mass both have the same properties; the magnitudes of both are independent of the galilean frames of reference and both are constants (with respect to time) of the system. I therefore identify one with the other; that is, I now define $m \propto 1/\ell$. Mass is now seen to be a geometric quantity. I then can write $mc\ell = \hbar/2$, where $h = 2\pi\hbar$, Planck's constant. That is, I can write this just because space and time intervals in the small are identical and mass is defined in terms of ℓ. Then $(\Delta x)(\Delta p) = mc\ell = \hbar/2$, and we see that the only way this result differs from the Heisenberg Uncertainty Principle is that the latter has both equal and unequal signs in its formulation, although the two equations express quite different physics.

50. I now can derive a de Broglie relationship from the random walk. Such a derivation is not required for the development of quantum mechanics but is exhibited to show that the de Broglie relationship arises not so much out of quantum mechanics but out of the statistics of the random walk.

Let $\Psi(x)$ be the amplitude of the probability function, $P(x)$. If $P(x)$ is the gaussian function we can then write $\Psi(x)$ as,

$$\Psi(x) = \frac{e^{(-x^2/4(\Delta x)^2)}}{(2\pi)^{1/4}(\Delta x)^{1/2}}$$

In evaluating its fourier transform,

$$\phi(k) = \frac{1}{\sqrt{2\pi}} \int_{-\infty}^{\infty} \Psi(x)e^{-ikx} \, dx$$

we find that

$$\phi(k) = \frac{e^{(-k^2/4(\Delta k)^2)}}{(2\pi)^{1/4}(\Delta x)^{1/2}}$$

Here $\Delta k = 1/(2(\Delta x))$, the standard deviation of k, so that $(\Delta x)(\Delta k) = \frac{1}{2}$. Since $(\Delta x)(\Delta p) = \hbar/2$ we infer (except for an arbitrary constant, which is taken as zero) $p = \hbar k$, the de Broglie relationship. *This* is the source of the so-called "wave" nature of matter in quantum mechanics—and it is not even quantum mechanical! It should be noted that this "wave" nature applies to an ensemble of objects or, if it does apply to a single object, it applies only over many (an infinity of) instants of time.

I conclude from this that the de Broglie relationship is not necessarily a quantum mechanical result, but is rather a consequence of a random walk distribution as presented in this chapter.

51. I now exhibit the derivation of the well-known result that the function, $\Psi(x)$, given above is a solution of the diffusion equation

$$\frac{\partial \Psi}{\partial t}(x,t) = a \frac{\partial^2 \Psi}{\partial x^2}(x,t)$$

Given the initial condition, $\Psi(x,0)$, the solution of the diffusion equation is given, by the separation of variables, as

$$\Psi(x,t) = \int_{-\infty}^{\infty} \phi(k)e^{-ak^2t}\, e^{ikx}\, dx$$

where

$$\phi(k) = \frac{1}{\sqrt{2\pi}} \int_{-\infty}^{\infty} \Psi(x,0)e^{-ikx}\, dx$$

Taking $\Psi(x,0)$ as the Gaussian function above at $t = 0$, so that $\Delta x = \Delta x_0$ and solving for $\phi(k)$ as was done above and substituting in the integral equation for $\Psi(x,t)$, we find

$$\Psi(x,t) = \frac{(\Delta x)^{\frac{1}{2}}\, e^{(-x^2/4(\Delta x)^2)}}{(2\pi)^{\frac{1}{4}}(\Delta x)^{\frac{1}{2}}}$$

where

$$(\Delta x)^2 = (\Delta x_0)^2 + at = \overline{(x_0 + x_1)^2}$$

However, since $\bar{x}_1 = 0$, then

$$(\Delta x)^2 = \overline{(x_0 + x_1)^2} = \overline{x_0^2} + \overline{2x_0x_1} + \overline{x_1^2} = (\Delta x_0)^2 + (\Delta x_1)^2$$

$$= (\Delta x_0)^2 + (\Delta v)^2 t^2$$

where $(\Delta v)^2 = c^2/n = c^2\tau/t = \ell c/t$, as I have already shown. Therefore, $(\Delta x)^2 = (\Delta x_0)^2 + \ell ct$ and $a = \ell c$, so that the diffusion constant is proportional to ℓ. Thus, as claimed, the random walk is simply a classical diffusion where the probability distribution function is

$$P(x) = |\Psi(x,t)|^2 = \frac{(\Delta x_0)\, e^{(-x^2/2\sigma^2)}}{\sqrt{2\pi}\,\sigma}, \qquad \sigma^2 = ((\Delta x)^2 = (\Delta x_0)^2 + at$$

52. To summarize:

In classical physics an object is defined as an analytic space function of time where space and time exist only where and when objects exist. By defining the object now as a random walk of a non-zero space-time *interval* it is possible to base the concepts of mass, space, and time on this interval.

We see now how the concept of object is the font for the concepts of mass, space, and time, how all of them arise from the concept of object. But in this achievement, we find that the object itself can no longer be defined in a finite time; the defining function is no longer an analytic, but rather a random walk function with a non-zero length step, a function that cannot be defined in a finite time. Thus, although we have gained the integration of the concept of mass into that of object, we have lost the basic ontology, that of a defined object. The problem is resolved in the next chapters, on quantum mechanics, although to some extent we may still be in trouble.

Furthermore, we see that an *ensemble* of classical objects defined by sufficiently small step lengths making a random walk is defined by a definite distribution function, $P(x,t)$. From this we can derive both an uncertainty principle $(\Delta x)(\Delta p) = \hbar/2$ and the de Broglie relation, $p = \hbar k$. Thus, both of these principles result from a classical random walk and are not the basis of quantum mechanics, although in some fashion yet to be determined, they also are *in* quantum mechanics.

Also, since in the random walk presented in the chapter on relativity there is no preference for the object to make a step in one direction more than in the opposite direction, it must be that $p = q \Rightarrow v = 0$ *always* for *all* objects. This, of course, is entirely unreasonable. This will be resolved in the next chapter on the quantum object.

VI

Quantum Object (Schroedinger)

53. As we saw in the last chapter, by integrating the concept of mass into that of object by identifying step lengths inversely with mass, we lost our basic ontology; no longer do we have the ability to state the defining function of the object at an instant of time. It seems that now indeed we are in a quandary: how can an object be defined both as a spatial analytic function of time and yet, in order to define a mass, consist of a series of discrete steps defining an interval: i.e., be a random walk? In an attempt to resolve these paradoxes, I return to the concept of the random walk object and see if there are implications in it that were overlooked.

54. A random walk object in time τ makes a step either $+\ell$ or $-\ell$. Since there is no preference for either $+\ell$ or $-\ell$, we conclude, as stated at the end of the last chapter, that in a long series of steps 50% will be $+\ell$ and 50% will be $-\ell$. That is, the probability of the object making a step to the left or right is the same, ½. But this inference from no preference is not valid: what *is* valid is that in 50% of the events, there is a preference for the object to take a step $+\ell$ and in 50% of the events there is a preference for the object to take a step $-\ell$. That is, the very fact that an object *takes* a step to the left or right *means* that it prefers—in fact, *must go* (because it *did* go)—in that direction. The meaning of the word, "preference" is not normally meant to imply certainty, but in this case, since there is only a single criteria for "preference"—whether or not a given event will or will not occur—it does imply certainty. If the resulting right or left step is certain, then that implies an analyticity, some unexpressed structure to the random walk, which of course does not exist. The only other possibility is the one already consid-

ered and rejected in the chapter on classical physics; i.e., the object disappears at some time and then another object appears at time τ later at a distance ±ℓ. Then, of course, objects consequently *would* appear equally to the left and right. However, since all these objects are quite independent of each other, each, so to speak, coming into and going out of the world on its own, there would be no reason for one to appear at a fixed time τ after one disappears and no reason for one to be at a fixed distance ± ℓ from the former one; that is, there is no reason for such a correlation. In fact, the probability for this to happen would be zero. Our original question then remains: how can an object make a non-preferential spatial step in a time τ? The answer is, of course, that it cannot, since ℓ = ±cτ. And yet, since it *cannot make no step,* what does it do?

55. It is here, in the answer to this question, in the resolution of this paradox, that we fortuitously turn our backs on classical physics and take the leap into quantum mechanics, from an object defined by either an analytic or random walk function to an entirely different kind of object. I realized that for the object to be truly non-preferential in its step, there was only one possibility, and that was that the object must go in *both directions*, to *both* +ℓ and − ℓ simultaneously. That is, we must give up the classical concept of an object, namely, that an object is defined by a spatial *function* of time. (Perhaps the writer who wrote about Buridan's Ass starving midway between two identical bales of hay had insight some of the rest of us did not yet have.[6]) Thus, it is not only that space is not a plenum in which objects swim about, but it is that, except for classical objects, objects do not even define space, that space does not otherwise exist. I will discuss this subject in more detail later on.

56. Non-preference means that an object takes a step to *both* the left *and* the right. Thus, the object makes a discrete walk with no disorder or randomness; at each step there is no preference. Such a walk, since non-preference is unique, is itself unique. Thus, since there is at each "step" no preference, there is only *one* function, $\Psi(x,t)$, describing the state of an object having a given initial state. Such a function, now, defines what

I call a *quantum object*. Such an object,in time τ, makes a step to *both* spatial locations +ℓ and −ℓ. That is, a quantum object at (x,t) = (0,0) will, at time t = τ be at both spatial locations +ℓ and −ℓ. At time t = nτ, the object will be simultaneously at points, x = ±jℓ, j ≤ n, j and n both even or both odd. Thus, for n = 1, the object is at both +ℓ and −ℓ; for n = 2, the object is at points, x = 0, +2ℓ and −2ℓ, all simultaneously. I wish to make clear that I do not mean by this that at n = 1 the object is half at each location or partly at each location or has a "probability" of 50% of being at each location. It is only at a measurement, the object then being in space, that it has a "probability" (of 50% in this case) of being at one or another location. When it is in "nonspace" it is *entirely* at *both* locations. At every value of t = nτ the object is at *all* the locations given by the above formula; i.e., spaced 2ℓ from each other. I will call the set of spatial locations or points of a quantum object its *nonspace*. That is, an object, except for a classical or random walk object, no longer exists in space but rather in nonspace.

In order to support the necessity for this quantum "non-hypothesis" even further I will further discuss the nature of a *classical* random walk. Such a random walk, as mentioned above, can be described correctly as one in which, as the number of steps approaches infinity, the fraction of positive and negative steps each approaches ½. It is confusing and vague, in light of the work done here, to say that there is "no preference" for a positive or negative step. On the contrary, one-half actually "prefer" making a positive step and one-half actually "prefer" making a negative step. The rationale for this claim is that such steps were actually made. That is, "preference" is a rather weak or inadequate term; the fact that each of the steps were made left or right was *necessary*. By necessity is meant determinate; that is, the fact that a step was made left or right was actually determined by the pre-existing conditions. This, of course, is exactly the situation with the throwing of dice—any game of chance—the motion of molecules in ideal gases, all events of a classical nature no matter how complicated. The concept of probability applied to classical events; e.g., a classical random walk, has nothing to do with any ontological uncertainty in the walk itself but only with the *lack of knowledge* of the determing factors of the walk. If these factors are sufficiently complex and varied, it is then not unreasonable to conclude that they will result in the binary event (step in a random walk) to go one way as many times as another. That is, the use of probability here is a *practical* matter, an *epistemological tool and not an ontological statement* and therefore is, if not seductively obscurant, certainly irrelevant.

However, is it possible that such a classical random walk can be non-determinant and "free"? Can each step truly be independent of its past, its history? Yes, of course, that is possible, but then not even a probability function can be defined for this kind of "classical" random walk. Therefore, not even the concept of "no preference," a 50-50 probability function, can be defined for such a walk. Therefore, if there is to be a "no preference" random walk and non-deterministic (which it must be), it cannot be a classical random walk, but a quantum random walk deterministic in nonspace, as described above.

Now if a measurement, which requires a classical object, is made on the nonspace of the object, then the result described above is totally non-determined except for the restrictions of possibilities determined by the state of the system (its various nonspace "positions") and the kind of measurement made; i.e., by the array of classical objects used. There are, otherwise, no inherent probabilities in going from one state to another.

It should be mentioned in this context that the state of an object is determined by initial conditions—its initial measurements; it has nothing to do with an ensemble of objects in the same state. One can *determine* the state of an object, however, by transforming an ensemble of objects all in the same state into space. The confusion of *defining* the state of a *single* object in terms of an ensemble is now removed.

I wish to emphasize that just as the identity of space and time "in the small" is the ontological source of special relativity, the "non-preference" walk described here is the ontological basis of quantum mechanics.

57. Thus, at least in the case given above, the original space point location completely determines its nonspace locations

(points in nonspace) at later times. More generally we shall see shortly that actually, not one, but two conditions are required. Furthermore, still having a step length defines the mass. It seems then, at least for this very simple case, that non-preference for a single object allows us once again to define the state of an object and also its mass. The question now is: Suppose that we are given *any* (arbitrary) initial state, then is the state at a later time still determined by the initial state? This is a "Green's function" or "kernel" type of question: is there a function that will transform or "carry" an initial condition into the future (or, for that matter, into other regions of space)? The answer is "yes" and the function turns out to be "no function;" that is, a function arising from no preference, no restrictions except for that arising from the requirement for a fixed step length or for an object to have a mass. The equation giving this determination in its first incarnation is called the *Schroedinger* equation and in its second incarnation is called the *Dirac* equation. This is how the attempt to define the concept of object coherently, a way that incorporates the non-principle of no preference, leads to the quantum object from which the above equations arise.

58. In summary, space, as stated in section 25, does not exist exterior to or outside an object, but is rather a *property* of a *classical* object. But, as we see, it is a property *only* of a classical object and not of a quantum object. Since a quantum object is at many locations at a given instant of time, these locations cannot be that of space; we say that these positions are positions or values of *nonspace* and for the moment refer to nonspace as a property of a *quantum* object. The concept of nonspace will be developed in the chapter on Nonspace and Measurement.

As with space, there is no time except where and when there is a classical object. Thus, if a quantum object is said to be at various nonspace locations *at a given time*, what is meant is that if the quantum object were at *one* of its nonspace locations (and thus a space location also) it would be there at a given time which would be the time of the classical object at that location. What is meant by the "time" of the quantum object when it is in nonspace is yet to be determined. The consequence of this distinction between the time of the classical object and that of the quantum object itself now leads, first, to the Schroedinger equation and then to the Dirac equation.

59. Suppose now that an object at time $t = t_0$ is put into a nonspace state having a gaussian distribution of positions. How this is to be done is described in the chapter on Nonspace and Measurement. In quantum mechanical texts it is simply assumed that this can be done—at least it is never questioned or the way it is to be done is never explained, as far as I know.

Assuming now that it can be done we then ask the question: If the standard deviation of the non-positions is Δx_0, what will the standard deviation, Δx, be at a time t_1 later, where $t = t_0 + t_1$? Recall now the clear distinction between space and nonspace, time and the "time" of the quantum object: the quantum object, not being a classical object, does not exist in space and time; but, as described above, nonspace is *defined*, as it must be, in terms of space, and later in this chapter it will be shown how to define the "time" of the object in terms of time (of the classical object).

60. In the chapter on the classical random walk object, the formula for the standard deviation is a function of the step length and the number of steps taken, $\Delta x = \sqrt{n}\, \ell$. I now use this formula to find the variance and standard deviation of the object at a later time. The first question that must be asked is; how many steps, n_1, have been taken in time t_1? Remember that in time t_1, no measurements are made, so that the object is in nonspace. If Δx *were* to be determined a time t_1 after the previous measurment then only a *single* step in *space* would have been taken, the object being in nonspace during this time. The argument for this conclusion is the same as for Δx being the standard deviation; if the quantum object's nonspace positions *were* to be transformed into a position in space (by a measurement, thus bringing the object into space), only *one* step in space would have been taken. Therefore, even though the object is taking steps in nonspace, in time t_1 it took only a single step in space, i.e., in between measurements.

61. If the question were to be asked: how many steps, n_2 were taken in time $t = t_2$, the answer would still be the same, namely only one since the object has been in nonspace during this time. Thus, no matter *what* the time between two instances of the quantum object, only one spatial step has been taken. This, of course, is not the number of steps taken in the nonspace. This and the time in nonspace will be determined in the derivation of the Dirac equation.

62. Once again, I wish to make clear that it is not that the object is actually brought into space at time t_1 later, but only that the standard deviation is defined as a *spatial* standard deviation. Thus, since the standard deviation is defined in terms of space at two different times, there is only a single *spatial* step between the two times. What remains in order to evaluate the standard deviation now is the spatial increase, which can no longer be simply ℓ. This is done in the following manner.

63. From the chapter on the classical random walk object, again, I have derived an uncertainty principle for the normal distribution function, $(\Delta x_0)(\Delta v_0) = \ell c$ where $\Delta x_0 = \ell \sqrt{n_0}$ and $\Delta v_0 = c/\sqrt{n_0}$, which I am taking as the initial nonspace specification, at $t = t_0$. From this, I get that $\Delta v_0 = \ell c/\Delta x_0 = \hbar/(2m(\Delta x_0))$.

Since, from the previous chapter

$$(\Delta x)^2 = (\Delta x_0)^2 + (\Delta x_1)^2 = (\Delta x_0)^2 + (\Delta v_1)^2 t_1^2$$

where, since

$$\Delta v_1 = \Delta v_0 , \qquad n_0 \gg 1$$

$$(\Delta x_1)^2 = (\Delta v_1)^2 t_1^2 = (\Delta v_0)^2 t_1^2 = \frac{\hbar^2 t_1^2}{4m^2(\Delta x_0)^2} .$$

we have

$$(\Delta x)^2 = (\Delta x_0)^2 + \frac{\hbar^2 t_1^2}{4m^2(\Delta x_0)^2} .$$

This, of course, is exactly the expression for the dispersion of a "wave packet"; that is, the variance of the nonspace of the object derived in standard texts by other means.[2] Furthermore, since Δv remains constant but Δx increases, we have $(\Delta x)(\Delta p) \geq \hbar/2$. This, of course, is the Heisenberg Uncertainty Principle.

64. In the classical random walk the standard deviation increases with $\sqrt{t}$; this is the way noise or randomness varies with time. In the nonspace walk, the standard deviation increases with t; this is the way information or uniformity varies with time. What happens when the object transforms from nonspace to space is another matter and will be discussed in the chapter on measurement.

In the classical random walk, the standard deviation in the velocity decreases with time so that it is determined by Δx; i.e., $\Delta v = \ell c / \Delta x = c / \sqrt{n}$. In the nonspace walk, Δv remains the same, thus producing the Heisenberg Uncertainty Principle, $(\Delta x)(\Delta p) \geq \hbar/2$.

The probability distribution function then for a nonspace random walk is

$$P(x,t) = \frac{e^{(-x^2/2(\Delta x)^2)}}{\sqrt{2\pi}\,\Delta x}, \qquad (\Delta x)^2 = (\Delta x_0)^2 + \frac{\hbar^2\, t^2}{4m(\Delta x_0)^2}$$

65. In order now to *derive* the Schroedinger equation,

$$\frac{\partial \Psi}{\partial t}(x,t) = \frac{i\hbar}{2m}\,\frac{\partial^2 \Psi}{\partial x^2}(x,t)$$

I look once more at the diffusion equation,

$$\frac{\partial \Psi}{\partial t}(x,t) = a\,\frac{\partial^2 \Psi}{\partial x^2}(x,t)$$

where a is the diffusion constant. In the preceding chapter it was shown that a gaussian function of the form

$$\Psi(x,t) = \frac{e^{(-x^2/4(\Delta x)^2)}}{(2\pi)^{1/4}(\Delta x)^{1/2}}$$

is a solution of the diffusion equation, where $(\Delta x)^2 = (\Delta x_0)^2 + a\,t$ and that if a is real, then

$$P(x,t) = |\Psi(x,t)|^2$$

66. Now, suppose that a is imaginary, or equivalently, that a is real and t is imaginary (it), so that the diffusion equation can be written as

$$\frac{\partial \Psi}{\partial t}(x,t) = i|a|\frac{\partial^2 \Psi}{\partial x^2}(x,t)$$

But $|a| = c\ell = mc\ell/m = \hbar/2m$, so that the diffusion equation becomes

$$\frac{\partial \Psi}{\partial t}(x,t) = \frac{i\hbar}{2m}\frac{\partial^2 \Psi}{\partial x^2}(x,t)$$

This, of course, is nothing but the Schroedinger equation; its solution for the given initial condition is nothing but the solution of the diffusion equation with a or t imaginary,

$$\Psi(x,t) = \frac{(\Delta x_0)^{\frac{1}{2}}\,e^{(-x^2/4(\Delta x)^2)}}{(2\pi)^{\frac{1}{4}}(\Delta x)^{\frac{1}{2}}}, \qquad (\Delta x)^2 = (\Delta x_0)^2 + i|a|t$$

The probability distribution function is then,

$$P(x,t) = \Psi^*(x,t)\,\Psi(x,t) = \frac{(\Delta x_0)}{\sqrt{2\pi}}\,\frac{e^{(-x^2/4[(\Delta x)^2])^*}}{[(\Delta x)^{\frac{1}{2}}]^*}\cdot\frac{e^{(-x^2/4(\Delta x)^2)}}{(\Delta x)^{\frac{1}{2}}}$$

where

$$[(\Delta x)^2]^* = (\Delta x_0)^2 - i|a|t$$

or

$$P(x,t) = \frac{1}{\sqrt{2\pi}\,\Delta x}\,e^{(-x^2/2(\Delta x)^2)}$$

where,

$$(\Delta x)^2 = (\Delta x_0)^2 + \frac{|a|^2 t^2}{(\Delta x_0)^2} = (\Delta x_0)^2 + \frac{\hbar^2 t^2}{4m^2(\Delta x_0)^2}$$

Thus, a random walk in nonspace—a single step in time—(starting with a gaussian probability distribution function) produces a gaussian distribution function which is exactly the same function of time that a random walk in imaginary time would produce; that is, *given* P(x,t) from the random walk in nonspace, then its amplitude, $\Psi(x,t)$, is a solution of the Schroedinger equation and the Schroedinger equation can be derived from it. The question now is: does the Schroedinger equation describe all such walks in nonspace no matter what the initial distribution is? That is, does the particular solution above, $\Psi(x,t)$, define a kernel or propagation function? The answer is "yes." This can be shown in the following manner.

67. The solution of the diffusion equation can be written in terms of the kernel or propagation function, $K(x - x',t)$, as

$$\Psi(x,t) = \int_{-\infty}^{\infty} K(x - x',t) \ \Psi(x',0) \ dx'$$

where $\Psi(x',0)$ is the amplitude at $t = 0$; i.e., the initial condition. The solution of the diffusion equation, however, can also be written, as I have previously indicated, as

$$\Psi(x,t) = \int_{-\infty}^{\infty} \phi(k) \ e^{-ak^2t} \ e^{ikx} \ dk$$

a, real or imaginary, where, once again, we write,

$$\phi(k) = \frac{1}{\sqrt{2\pi}} \int_{-\infty}^{\infty} \Psi(x',0) \ e^{-ikx} \ dx'$$

Then, by substitution and manipulation, we have

$$\Psi(x,t) = \int_{-\infty}^{\infty} \Psi(x',0) \left[\frac{1}{\sqrt{2\pi}} \int_{-\infty}^{\infty} e^{-ak^2t} \ e^{ik(x-x')} \ dk \right] dx'$$

Therefore,

$$K(x - x',t) = \frac{1}{\sqrt{2\pi}} \int_{-\infty}^{\infty} e^{-ak^2t}\ e^{ik(x-x')}\ dk \sim e^{(im(x-x')^2/2\hbar t)}$$

and since,

$$\frac{\partial \Psi}{\partial t}(x,t) = \int_{-\infty}^{\infty} \frac{\partial K}{\partial t}(x - x',t)\ \Psi(x',0)\ dx$$

$$a\frac{\partial^2 \Psi}{\partial x^2}(x,t) = \int_{-\infty}^{\infty} a\frac{\partial^2 K}{\partial x^2}(x - x',t)\ \Psi(x',0)\ dx$$

and,

$$\frac{\partial K}{\partial t}(x - x',t) = \frac{1}{\sqrt{2\pi}} \int_{-\infty}^{\infty} (-ak^2)\ e^{-ak^2t}\ e^{ik(x-x')}\ dx$$

$$a\frac{\partial^2 K}{\partial x^2}(x - x',t) = \frac{1}{\sqrt{2\pi}} \int_{-\infty}^{\infty} (-ak^2)\ e^{-ak^2t}\ e^{ik(x-x')}\ dx$$

Then,

$$\frac{\partial \Psi}{\partial \tau}(x,t) = \frac{i\hbar}{2m}\ \frac{\partial^2 \Psi}{\partial x^2}(x,t)$$

for *all* initial conditions, $\Psi(x',0)$.

Thus, from an object that is a fixed length random walk in nonspace and imaginary time, we can derive the Schroedinger equation. I conclude, therefore, that quantum mechanics, at least non-relativistic quantum mechanics, is the description of an object making a "no-preference" walk in nonspace and imaginary time, while making only a single step between two (real) time instants. Determinism exists, but only in nonspace and only between two time instants. In fact, nothing at all has been assumed about the behavior of objects *in* nonspace and imaginary time; any property of the object between two time instants, such as having an imaginary diffusion constant, or, equivalently, existing in imaginary time, is a consequence of the nature of the object *at* (real) time instants. What happens in nonspace is explicitly spelled out in the next chapter on the Dirac equation.

Before going on to the Dirac equation, it should be noted that the second paradox of a classical random walk, discussed in the chapter on relativity, namely, that v should always be zero, is now resolved. *That which determines the v of an object is the initial state or measurements on the nonspace.* This, of course, is exactly what happens to an object making a path in a bubble or cloud chamber; that is, there is a continual measurement on it by its immediate environment.

VII

Quantum Object (Dirac)

68. In the last pages I found that the necessity of non-preference for a random walk object of a non-zero step length led to an object having the property of nonspace and imaginary time. This conclusion led directly to the Schroedinger equation. The only assumption made was that the random walk of the object could be approximated by a gaussian distribution. But, as indicated on page 51, this approximation holds only when s^3/n^2 $(= xv^2/\ell c^2) \to 0^5$. Thus, if v is not close to zero, the derviation no longer holds and a more basic approach is required. The approach developed here requires an expansion of the previous analysis which will allow all velocities to exist, since otherwise the relativistic nature of the object would not be revealed. Eventually this approach results in the Dirac equation.

More fundamentally, however, because the step length, ℓ, is fixed, the positions an object can occupy at a measurement are discrete, thus raising the question of what would happen if a measurement were made *not* at the possible positions of the object or *not* at the time instants in which the object can appear. (In fact, since τ and ℓ are instants and positions, the probability of a measurement event ocurring at *any* given position and time is zero.) The problem to be solved then is to define an object in its nonspace as truly having, at *all* instants of time, "all possibilities," thus maintaining the quantum "non-hypothesis," and yet to be able to define the concept of mass. Such a definition, therefore, must be a restriction on "all possibilities" of nonspace. Thus, I envisage a typical path in nonspace and imaginary time to consist of random walks of a distribution of lengths based on the concept of non-preference.

69. Let us, therefore, start over, without making the above approximation. Again, the basic "non-hypothesis" is that an object is defined by a non-preferential random walk. This has been shown in the previous chapter to lead to the conclusion that the "spread" of the standard deviation of the increase in the nonspace is proportional to the time. The probability distribution defining the nonspace at any time is determined only by the initial conditions that established the nonspace and is not determined by what is going on in the nonspace, where the object is making a random walk of all possibilities in imaginary time. Suppose now we ask: in time t and distance $\Delta x = 0$, what are the possible "motions" in nonspace and imaginary time?

Once again I start off with a random walk having a fixed step-length, ℓ, and time interval, τ, which, however, I eventually let approach zero. Now, since the probabilities of going left or right are equal, it is possible to determine the probability of a reversal in N steps and the average number of reversals, $\overline{R}$, in N steps say for $\Delta x = 0$ in time, t. Then, the probability of a reversal in N steps in imaginary time is $P = i\overline{R}/N = (\overline{R}/t)(it/N) = (\overline{R}/t)(i\in N/N) = (\overline{R}/t)(i\in)$, where $\in$ is the time between steps in the nonspace. [t is real time, time defined by measurements. $i\in$ refers to imaginary time intervals between measurements.] Now from the previous chapter we know that the standard deviation of the increase in the nonspace is proportional to the time of "expansion" of the nonspace. However, the standard deviation is also proportional to the average number of reversals, $\overline{R}$. Thus, the average number of reversals per unit time must be a constant; i.e., $\overline{R}/t = 1/t_0$, a constant, where t_0 is the average time between reversals. A sample calculation shows that for an electron at "rest" there are about 10^{20} reversals per second. This is what is known as the "zitterbewegung." Thus, $P = i\overline{R}/N = i\in/t_0$, and the probability of a reversal in N steps is equivalent to the probability of a reversal in any time, $\in < t_0$. The probability of R reversals in any such path is then given by $P(R) = (i\in/t_0)^R$.

70. Now t_0 is an arbitrary parameter not dependent at all on the initial conditions or the kinematics of the motion of the

object. Furthermore, since mass is not yet part of the Dirac object theory, it can be defined in terms of t_0, $m = \hbar/2c^2t_0$. Thus, I have derived an expression for P(R), what Feynman calls the path contribution to the amplitude, K(x,t), of the Dirac equation.

If the total number of "paths" with R reversals in going between the beginning and end position points in time t is designated as $\phi(R)$, then the total number of paths in going this distance in time t, is $K'(x,t) = \sum_R \phi(R)\, (i\epsilon/t_0)^R$. By taking the limit of $dK'/d\epsilon$ as $\epsilon \rightarrow 0$, I get the "amplitude" per unit time, K(x,t). It is in this manner that I derive the fact that every walk in nonspace is a continuous walk interrupted by random reversals at every point of the nonspace defined by K(x,t).

This expression for K(x,t), however, turns out to be the propagator or Green's function for the Dirac equation! That is, in the attempt to develop a coherent concept of object, I have derived the Dirac equation. [Feynman, with his amazing insights, presented this formula for K(x,t); however, he did not give a derivation for the propagator. My derivation, based on "non-préference," gives identical results to those Feynman got from his variational approach.[7]] The actual calculation of the value of K(x,t) was done by Jacobson and Schulman. The calculation in terms of the momentum, K(p,t), was worked out somewhat earlier by Gersch. These formulas are:

$$K(x,t) = \frac{mc^2}{\hbar}\,[iJ_0(z) - \gamma J_1(z)], \qquad z = \frac{mc^2 t}{\hbar\gamma}, \quad \gamma = \left[1 - \frac{v^2}{c^2}\right]^{-\frac{1}{2}}$$

and

$$K(p,t) = \frac{1}{2}\left[1 + \frac{mc^2}{\hbar}\right]e^{\frac{-iEt}{\hbar}} + \frac{1}{2}\left[1 - \frac{mc^2}{\hbar}\right]e^{\frac{iEt}{\hbar}}$$

Gersch has also shown that except for the phase factor,

$$K(x,t) \sim e^{imx^2/2\hbar t},$$

so that the propagator of the Schroedinger equation turns out to be exactly as anticipated by Feynman, which I derived in the last chapter.[8, 9]

K(x,t) is a "probability" distribution function in nonspace and imaginary time: we see that it is complex. It is therefore interpreted as the *amplitude* of the probability distribution function in real time and space. However, the probability distribution functions are not the square of the amplitude function, as in the case of the Schroedinger equation, but since the amplitude function already is a "probability" function (albeit complex), the probability density functions, $|K(x,t)|$ and $|K(p,t)|$ are the magnitude of the amplitudes, so that

$$|K(x,t)| = \frac{mc^2}{\hbar}\left[J_0^2(z) + \gamma^2 J_1^2(z)\right]^{\frac{1}{2}}$$

and

$$|K(p,t)| = \frac{1}{\sqrt{2\pi}}\left[1 + \frac{m^2c^4}{E^2} + \left(\frac{pc}{E}\right)^2 \cos\frac{2Et}{\hbar}\right]^{\frac{1}{2}}$$

These distributions, however, result from the superposition of both positive and negative energy functions. For only positive and only negative energies, we have,

$$K(p,t) = \tfrac{1}{2}\left[1 + \frac{mc^2}{E}\right] e^{(-iEt/\hbar)}$$

and

$$K(p,t) = \tfrac{1}{2}\left[1 - \frac{mc^2}{E}\right] e^{(iEt/\hbar)}$$

so that

$$|K(p,t)| = \tfrac{1}{2}\left[1 \pm \frac{mc^2}{E}\right]$$

is flat only for small v. We see that even for single energy values, we do not have a flat distribution generally. The existence of an upper bound to the velocities prevents the distribution from being flat, as it is for the Schroedinger distribution.

71. An object satisfying the above simple requirements is called a Dirac object. At $t = 0$, $|K(x,t)| = 1/2t_0$ so that the reversal time, and thus the mass, can be determined at an instant of time. Nevertheless, since $|K(x,t)|$, a probability function, is being determined, an infinity of measurements are required at a single instant of time. What this means for the ontology of an object will be discussed in a later chapter. Of course the propagator function, $|K(x,t)|$, is also determined, as is the state function, $\Psi(x,t)$, once the initial function, $\Psi(x,0)$, is established; that is, $\Psi(x,t)$ is determined until, as will be shown in the next chapter, a measurement is made. The Dirac object is the most developed concept of object I deal with in this work. Thus, it is interesting to compare and contrast the Dirac object with the classical object. The classical object, being defined by an analytic function, was defined at a space-time point by its derivatives; its mass was defined by another procedure entirely; e.g., by its resistance to a change in its motion or by the stretching of a spring. The Dirac object, on the other hand, is defined, kinematically, simply by a *single* measurement, $\Psi(x,0) = \delta(x-x_0)$; its mass, however, is defined *also* at that instant of time ($t = 0$), by $|K(x,t)|$, a probability or distribution per unit time $2(mc^2/\hbar)$, defined by the objects at that location at that time. However, the completion of this new ontology requires a statement of the concept of measurement and a further exegesis of these results.

VIII

Nonspace and Measurement

72. I shall now develop further the meaning of the concept of nonspace in order to relate the concept of measurement to the concept of nonspace.

Points in nonspace are points or positions of, or in, *nonspace*, not points or positions of, or in, space. Thus, wherever there is a point in, or of, space, there is a classical object and wherever there is a classical object, it is said to be at a point of space. Space is said to be a *property* of a classical object. On the other hand, a quantum object—or simply, object—does not have the property of space; it has the property of nonspace; that is, it is at one or more positions or locations, not in space, but rather in nonspace. (This formulation will shortly be reworded more accurately.) By position or location in nonspace is meant that if these locations were *space* positions or locations, the (quantum) object would be at one of them. Furthermore, if classical objects were at *each* of these nonspace locations, the quantum object would be at *one* of these locations.

73. Recall now the nature of an object, any object, at a single space point at a given time; it will "move" in a totally non-preferential way. That is, as time increases, the object will find itself, at any given time, at all possible locations. The only restriction on these locations is determined by the mass of the object, which determines the average time between reversals of the random walk of the object (recall that the fact that the velocity is bounded is not really a restriction on the object's possibilities, but that it would not even be possible to define an object unless its displacement in a given time were bounded). Thus, for small masses the average time of reversal is large while for large masses the average time of reversal is small. If the mass is sufficiently large (let's arbitrarily say larger than

10^{-15} kg), so that it might, for our purposes, be considered infinite, then the average time between reversals is essentially zero. This means that all the possible positions or locations occupied by an infinite mass object (having zero velocity) is within any neighborhood, Δx, of its original position at any later time. I can now redefine the concept of classical object to be such an object; that is, it is a quantum object of essentially infinite mass. Since all non-interacting objects of the universe must be, unless there is a special situation, non-preferential, the redefined classical object, since it is of "infinite" mass, is then indeed a special kind of a (quantum) object, and space is a special kind of nonspace. Classical objects having zero velocity thus remain at the single, same location for all subsequent time, or if moving, have a trajectory; objects other than classical objects (quantum objects) will be at a continually increasing set of continuous positions or locations in nonspace. It is clear that at a later time a quantum object originally at a single point in space will occupy many locations or positions; is it possible for an object to transform from its non-space state into a space state? In fact, more generally, is it possible for it to go from one nonspace state into generally any other nonspace state at an instant of time? The answer to both of these questions is "yes." In order to understand how this happens I propose the following analysis.

74. Let us consider the first kind of random walk in nonspace we discussed, that which was the basis for the Schroedinger equation. That walk consisted of discrete steps, $\tau > 0$, in both directions. Again, if the object is originally at a single point in space, then at each step it shows no preference to go one way or another in its next step. (The resulting probability distribution, as has been shown in the derivation of both the Schroedinger and Dirac equations is not that of classical physics, the binomial distribution, but is rather flat or near flat.) That is, *no choice* is made, there are no restrictions except that of the step size, τ, or, equivalently, ℓ. On the other hand, suppose that a classical object before time τ were to be placed at either $+\ell$ or $-\ell$. Since a classical object is present at one of these locations, then that location, as the other must also be, is a location in space. Thus the object, which otherwise would have

gone to both locations, $+\ell$ and $-\ell$, and thus be in nonspace, no longer has this option, since both of these locations are now spatial locations. Thus, the object must go to just *one* of these locations and is itself still in space, as it was originally, at zero time. Thus, the presence of a classical object, in that it can specify a spatial location, *restricts* the possibilities of a quantum object and, so to speak, forces it to make a choice—that is, to go one way or the other. That is, the presence of classical objects, and thus space, create a preference or restriction for a quantum object, thus "forcing" it to go to only one of its two possible locations. If this procedure were repeated every time interval, τ, then the object would be making a random walk in space. Such a walk would be *truly* random, in that the way an object would go at each step would not be determined by any spatial pre-existing situation. That is, the presence of a classical object restricts the spatial possibilities of the object to *one* location, but cannot produce any further restrictions. In fact, the very meaning of nonspace itself is characterized in terms of what the spatial results would be *if* the object's nonspace positions were to become spatial positions. As shown in the chapter on the random walk classical object, any attempt to determine which direction the object will "choose" or be determined to go, even if the step time is zero, the presence of more than a single object results in a loss of the identity of the object and thus undermines the possibility of determinism, since determinism in space-time is based on that of the continuing identity of the object. *All* motions of quantum objects in *space*, random or apparently deterministic, result from the presence of classical objects "compelling" (quantum) objects to make spatial choices. Thus, space can be thought of as the secular purgatory of objects that have been removed from an enduring identity in nonspace. Compelled to be "deterministic" there, if you will, almost totally uncertain as to what their transformed condition will be, bereft of their past, they are eager to get back into nonspace again, even though their previous nonspace ancestry no longer exists.

75. I have shown that the "cause" of the indeterminism or acausality neither exists nor results from hidden forces, influences, or interactions, themselves classical concepts, but is sim-

ply inherent in the "atomistic" nature of space-time (although not of nonspace and imaginary time), each instant largely isolated and independent of all other instants. Past, present, and future all seem to be in different hardly approachable rooms. To ask *why* the object in space goes one way instead of the other is already to presume the wrong answer, i.e., that the object always exists in space as a classical object, that there is no nonspace. Let me now extend the analysis.

76. Consider the above object in nonspace after a number of steps. Suppose that a classical object is placed at one of the locations, x, that the quantum object would be present at after its next step; then the quantum object will not be able to be present at all of its positions in nonspace, since one of them is now the spatial position, x. The object can be either at all the other nonspace locations or it can be at only x, the spatial location. That is, if there are ten nonspace locations, and one space location, then one out of eleven times the object will find itself at x, the spatial location; ten out of eleven times it will have been transformed to the nonspace defined by the other ten locations. If there are two classical objects present, then two out of eleven times, the object will find itself at one of the spatial positions and nine out of eleven times it will have been transformed to the nonspace defined by the nine other non-spatial locations. If there are a thousand space locations and two nonspace locations then, if the object is not at one of the thousand spatial locations, it will be at the location of the two non-space locations. This is exactly the situation of an object going through two slits or holes (the two nonspace locations) without dividing. If the object found itself at the spatial location, x, or, for a single slit screen (only one nonspace location), at the slit (the single position), then there is a coincidence with a classical object or, for the latter, a unique *lack* of a classical object. This is known as a *measurement*, particularly a position measurement. The classical object is called the measuring agent.

77. As shown in Chapters VI and VII, quantum mechanical equations, such as the Schroedinger and Dirac equations, hold only in between measurements and have absolutely nothing to say about the result of a measurement itself. The Schroedinger and Dirac equations are based on and result from the "quan-

tum non-hypothesis," the fact that there is a single "step" in space between measurements, although there are many steps in nonspace and imaginary time which is the reality between measurements. Thus, if a (quantum) object is to confront a classical array which it has a 50% chance of hitting (say the classical array takes up 50% of the future nonspace positions of the object), whether or not it does hit it is not determinable by the wave function of the object. What is certain is that the presence of the classical array will result in a measurement; that is, the effect will either be at a single location at one of the classical objects of the array or it will not. Thus, we conclude that the wave function of the object after the measurement will have changed from what it was before the measurement and will correspond to one or the other of these two possibilities. Thus, even if the object does not "hit" or coincide with any position of the classical array (and thus there certainly is no interaction), the wave function is no longer the original wave function: *the nonspace of the object is no longer the same even though no interaction has taken place, but a measurement (on the nonspace) and thus a transformation (of the nonspace) has taken place.* What *can* be determined is the *probability* of either transformation taking place. Clearly, classical physics is *not* the basis of quantum mechanics but rather arises, as has been shown, from quantum mechanics under specialized circumstances. Thus, if the system considered is no longer just the object but the object and the classical array, measurement will take place within the system and thus *no quantum mechanical equation can describe the state function of the combined system over its history. This is not a matter of the incompleteness of the quantum mechanics but rather a necessary statement that the ontology of physics is not classical.* The significance of this conclusion, resulting from the clear distinction between measurement and interaction, cannot be sufficiently stressed: the reality of the world is such that there is no determinism, not even a "*probalistic determinism.*"

Since the above example is essentially identical to Schroedinger's cat paradox, the explanation given in this work resolves that paradox.

78. It is necessary to state the distinction between the wave function, $\Psi(x,t)$, and the nonspace and also to show their rela-

tionship. The wave function, as indicated in the introduction in the quote from Schiff's text, is in terms of the number of *measurements* giving a particular result. Nowhere in the explication of quantum mechanics (except in Bohm s deterministic attempt) is there any attempt to state the source or the ontology of these results. Nonspace and imaginary time, as developed in this work, is the reality from which the results of these measurements, and thus the wave function, $\Psi(x,t)$, arises.

79. Sometimes it is stated that a measurement "collapses" a wave function (here, nonspace) "instantaneously," thus resulting in "superluminal velocities." Since "instantaneously" means a very large or infinite *velocity* and velocity is defined as the distance traveled per unit time (for $\Delta t \to 0$), we obviously have a non-sequiter here since nonspace is not space and therefore there is no (spatial) distance being traveled or collapsed. Such a conclusion could only arise from a classical or deterministic ontology.

80. A more significant problem concerning measurement often mentioned in the literature is why it is only classical objects, apparently, which result in these sharp, non-predictable changes in the wave function, even though they are themselves only a special case of quantum objects.

My answer to this question follows Bohr essentially: all the determinations and measurements of the world are made by classical objects; i.e., objects that are relatively massive objects ($>10^{-15}$ kg). Therefore, since all concepts of reality, in order to have meaning, must be expressed in terms of measurement, they must be expressed in terms of classical objects and classical concepts.

81. However, the question has an even deeper significance because of the inconsistency in the nature of classical objects. A classical object, as considered in this work, exists, strictly speaking, only at a given location; that is, it is a point object. It is only because of this property of a classical object that we can even talk about a *coincidence of positions*, and thus define what a measurement is. However, a point object, as shown in this work, must have an infinite mass, an obvious impossibility. Measurements, however, are made with classical objects. That is, measurements are made with objects of sufficiently large mass so

that they can be said to effectively have a position. From the Uncertainty Principle, $(\Delta x)(\Delta v) \sim \hbar/m$, we estimate that for $\Delta x \sim 10^{-13}$ mm it would take over a hundred years for the nonspace to increase to 10^{-3} mm for a mass of about 10^{-15} kg. Thus, for any measurements we make, we can say that our classical objects are of effectively infinite mass and in space. Nevertheless, they are not in actuality. Therefore, the possibility arises that perhaps other than classical objects may make measurements. Perhaps there can be a coincidence of non-classical objects. Perhaps, since strictly speaking no objects, not even the objects we are considering classical, are classical, measurement is not a coincidence of positions but of nonspace "positions" or states.

I indicate here in a very preliminary way how this might occur, at least conceptually. To do this, I go back to the original concept of nonspace, defined in sections 55ff, as a non-preferential walk with a constant step-length, ℓ, in both directions, so that in time τ, the object's state function transforms, $(0,0) \rightarrow (\ell,\tau) + (-\ell,\tau)$, or in wave function notation, $\Psi(0,0) \rightarrow \Psi(\ell,\tau) + \Psi(-\ell,\tau)$. Since the state of a (classical) measuring instrument is defined by $\Psi(x) = \delta(x-x_i)$ or an array or collection of such objects, a measurement will result in either a coincidence with a position value of the object s nonspace, $x = x_i$, or a superposition of nonspace positions $\phi(x) = \sum_j \delta(x - x_j)$, where $j \neq i$. On the other hand, the measuring instrument need not be defined by classical objects, each at some position, $x = x_i$; it may be defined by a non-classical object such as $\Psi(x,t) = \delta(x-\ell,\tau) + \delta(x+\ell,\tau)$ or, more generally, by *any* state vector of a (quantum) object. A measurement done by such an object then is the coincidence of the "two-position" of the measuring instrument with the identical "two-position" of the nonspace of the (quantum) object being measured. Since the state function of an object is expressible as a linear superposition of any set of orthonormal vectors of the Hilbert space, then it is expressible in terms of a "two-position" or, for that matter, of an "n-position" state function; thus, this or similar kinds of non-classical measurements perhaps may occur in the confrontation of the nonspaces of (quantum) objects. Thus, perhaps measurement may be of a much more general nature than simply *classical* measurement; i.e., measurement made only with classical objects. The existence of interactions, from which we infer the fact that a measurement took place, perhaps need not arise only during classical measurements.

We conclude, therefore, that just as in the case where classical objects confront (quantum) objects, quantum mechanical equations no longer apply, it would also be true if (quantum) objects confront each other. It is in this manner that perhaps the concept of measurement may be generalized without the need for classical objects. Thus, it is not to be expected that quantum objects in their confrontation would obey any quantum mechanical law over their history.

82. Now an object in a nonspace may confront an array of classical objects. Then, as stated above, the object may either find itself at the location of one of the classical objects, or it may not. If it doesn't, then it is defined by the remaining positions of its nonspace. Thus, by the use of arrays of classical objects, *any* nonspace state of an object can be established. Every transformation of the state of an object into another nonspace or space state is simultaneously the elimination of the past of the object, and the establishment of an initial condition for the ensuing nonspace. If we wish, therefore, we can establish, as indicated in the chapter on the Schroedinger equation, a gaussian state function for an object or any other initial condition. (Generally, since both $|\Psi(x,t)|^2$ and $\partial/\partial t\,|\Psi(x,t)|^2$ are required, two such distributions at slightly different times will have to be established.[10])

83. By this time, it should have become fairly obvious that a measurement is *not* an interaction; interactions have a limited speed (c) of interaction while the concept of speed, being a concept based on space and time, does not apply to nonspace or its "collapse"; to do so will result in such non-sequiters as "superluminal velocities."

84. Some important ideas can now be restated:

By measurement is meant, first of all, the position coincidence of that which is being measured with a classical object. *All* measurements are either of position or interval—there are no other kinds, although in section 81 I suggested an extension of the concept of measuring agent to quantum objects.

Nevertheless, all non-classical phenomena must be and are eventually defined in terms of classical concepts, which in all cases eventually derive from the concept of measurement by classical objects, that is, coincidences of position, or of the distance or interval between two such coincidences.

Position measurement is defined as the position coincidence of that which is to be measured, classical or quantum, and the measuring agent, which is a classical object.

If classical objects don't exist, as they probably didn't at the beginning of the universe, then neither did space or time; there was only nonspace and imaginary time. Mass, however, could exist, since it is defined by the average reversal time of the random walk in nonspace and imaginary time.

If an object defined by a given nonspace state confronts an array of classical objects and if that object then does not appear at a spatial position so that it is at all other locations in nonspace, the state of the object has been transformed into a new state, one defined by these other locations. Clearly in this case there is a measurement, since the state of the object is no longer the same as it was before. Yet, there is no interaction. Exactly the same thing can happen if the object does appear at a single space position. There is no interaction with an object going through or coinciding with a hole in a screen. I conclude that measurement does not necessarily entail interaction.

Nonspace and imaginary time are just as "real" as space and time. In fact, space and time, in the last analysis, since classical objects are only a sub-species of quantum objects, exist only as a "limited" reality for classical, large mass objects; the basic reality of objects is nonspace and imaginary time. It is especially significant to realize, therefore, that the state of an object is transformed by classical objects and *not* by interactions or forces.

Measurement and interactions are distinct and separate realities, one having nothing to do with the other. Interactions, however, do appear to be involved in most measurements or position coincidences. This is because the measurement, or position coincidence, although momentary, the object then immediately going back into a nonspace state is, in most cases, compelled to remain at its position of coincidence by interaction. The result of short range attractive interaction is normally to *bind* the object being measured to the measuring agent, the classical object, by reducing the energy of the (quantum) object or by transferring it to the coincident classical object. (Long range or scattering forces are not dealt with in this work.) Thus, if the position of an electron is measured, say, by a coincidence with the position of a classical object such as an atom bound to other atoms, the electron then also interacts with that atom, used for measurement. By relieving the electron of a sufficient amount of its energy, the atom binds it to itself, either keeping it in a position state, or, by giving off its energy as a photon or using its energy for a chemical reaction, all at that position. However, even if there is no interaction, there can nevertheless still be measurement.

85. Both classical and quantum physics consist of laws (deterministic behavior) between measurements. The classical laws deal with the space-time behavior of an object; the quantum mechanical "laws" deal with the nonspace-imaginary time behavior of an object. Recall that a measurement (by a classical object) is a coincidence of the position of a classical object with the position of an object or the lack of such a coincidence (or coincidence of intervals); there may or may not be an interaction.

In classical physics, the establishment of initial conditions on a system A by a system B—the measuring instrument or agent—is also achieved by position coincidence and thus is a measurement. As stated previously, a measurement is not an interaction; in classical physics, if there is a measurement but no interaction then system A is not affected at all by the measurement. That is, system A remains a closed, self-determined system; the measurement, therefore, does not create an initial condition. On the other hand, if the measurement is accompanied by an interaction, then system A is no longer isolated, and from then on is no longer uniquely determined by its own past history. However, the *combined* system of A and B is not only self-contained but its future is determined by its past, no matter how complicated the interaction between A and B is during the measurement. In the case of a quantum object, the effect of interaction, between measurements, is also always deterministic, but in nonspace and imaginary time. But the process of measurement on a quantum object by a classical object (or perhaps even by another quantum object) may be viewed as the establishment of initial conditions. However, in this case, the combined system of object and measuring agent, although self-contained, as stated previously, is no longer deterministic. This, of course, holds whether or not one of the objects is a classical object. Any two objects, of which at least one is non-classical—or, more accurately, any two nonspaces—confronting each other are "measuring" each other and thus reality is never deterministic. The best that we can say is that any actuality defines its potentialities; the transformation of an actuality into a potentiality is the nature of change.

Measurement has nothing to do with "observation," "consciousness," "psychic or brain processes," etc. It is not fundamentally a *human* process; it is simply a "confrontation" of nonspaces and in one very important case, a confrontation of a nonspace with a very special kind of nonspace—a classical object.

IX

Summary and Exegesis

86. I am now ready to answer the question, "What is an object?" At first, however, this question is only for a *single* object.

a. An object is exactly nonspace restricted by a given value of t_0, the average "reversal" time in the nonspace.

b. *Nonspace* is the set of all nonspace paths between nonspace points, $(0,0)$ and (x,t); more generally, from any set of nonspace points at a given time to an other set of nonspace points at some other time.

c. *Nonspace paths* are paths where the velocities vary randomly along each path as either $+ic$ or $-ic$, where c is a constant.

d. The average number of discontinuities—or reversals—per unit time, $\overline{R}/t = 1/t_0$, is fixed; that is, it is fixed for *any* state of an object of the same mass. Therefore, mass is *defined* by t_0, the average time between reversals for zero velocity of the object. The fact that $\overline{R}/t = 1/t_0$ is fixed is a restriction on all path possibilities in the nonspace. Thus, mass, which is determined by t_0, is determined by the average of the reversal times of *all* paths in nonspace for zero distance moved between measurements.

e. An object therefore is *all* kinematical possibilities in nonspace and imaginary time, subject, however, to the restriction noted above, which gives it the property of mass.

f. The *state* of a nonspace is the function, $\Psi(x,t)$. This is determined (measured) by the probability density, $|\Psi(x,t)|^2$, and the time derivative of the probability density, $(\partial/\partial t)\,|\Psi(x,t)|^2$, at a given time.[10]

The function, $\Psi(x,t)$, is determined by measurement; that is, it is determined by arrays of classical objects. The measurement itself is the establishment of the initial conditions of the nonspace of the object. Each measurement produces a given initial condition, and, thus, a different state of the nonspace.

g. Measurement, although a restriction on nonspace, is not a "complete" restriction on it and, therefore, the result of the measurement is not classically deterministic. Thus, the measurement, say by a double slit diaphragm, may result in the object being either at the nonspace of both slits or at a single position of the space at the solid part of the two-slit diaphragm. Such an "incompleteness" is logically necessary to avoid the incoherence of classical determinism; that is, the theory now *is* complete.

h. Nonspace is defined eventually in terms of *intervals* (distances) between reversals, whose domain is the set of reals. The randomness of the length of these intervals in nonspace arises directly out of the lack of preference for any "path" or preferential procedure. Where only a single object is present, the restriction on all possible paths is the existence of an average time between reversals, which defines the mass of an object.

i. Thus, nonspace is the basic reality from which, by measurement, space, time, and classical objects arise.

j. However, as shown in Chapter VII, it requires an infinity (very many) of measurements in order to determine the average time between reversals, t_0, and thus the mass of the object.[11] This can be done on a single-object nonspace by *repeated* measurements over a "long" time interval or at a given instant of time on an ensemble.

k. On the other hand, no average time between reversals, t_0, may exist. The restriction on the nonspace may not be that the nonspace is equivalent to a *single* object; *many* objects may arise from it. Thus, it is the concept of nonspace and not that of object that is ultimately the basic ontology.

l. Frames of reference require classical or relativistic objects to define them and are not at the basis of quantum theory; they are purely a classical concept and thus arise in quantum mechanics only when measurements are made. In this context it is of interest to refer to the statement made by Dirac on p. 252 of his text, *The Principles of Quantum Mechanics*, 3rd ed., in which he states: ". . . there does not seem to be any natural way of generalizing this notion of an observable to make it cease to refer to a particular Lorentz frame." Could the source of this question reside in the fact that only in classical physics are there frames of reference?

X

Ontology I

87. In the preceding pages I have developed the concept of object and the even more basic concept of nonspace. In the course of doing so, principles and concepts basic to these concepts have been discovered.

This is what I have discovered.

88. The world, whatever it is, is "all possibilities" subject to certain restrictions. In being what it is; i.e., all possibilities, there are no preferences unless they arise out of the being of the world itself or that which is "outside" the world. How these restrictions or preferences arise is not generally understood. But the only meaning the words "possibilities" and "preferences" can have, must arise out of the way in which the world is known, contacted, identified, gotten into. The word for this identification is "measurement."

Measurement, of course, is only a measurement of what is, the world. What is the world? The world is nonspace. In the analyses done above it has been shown that nonspace subject to the restriction of having a definite value of the average time between reversals is identified with the concept of a single object.

Beyond this, however, nonspace may, when measured, produce more than a single object. One cannot say then, even with the mass restriction, that nonspace is objects or consists of objects, since there is no way to identify, distinguish, or even count objects in nonspace; objects exist only in space and time and are produced or *created* by a measurement. Thus, a further restriction on nonspace is required: that is, nonspace is also restricted by the *number* of objects a measurement can produce. That is, nonspace is restricted by a set of reversal times, t_i, each having a populations, n_i. It is clear then that the concept of object loses its status as the fundamental reality if nonspace is the source of more than a single object. We now can redefine the concept of object as a mass existing in space and time; its mass requires the measurement of a distribution of positions, $|K(x,t)|$, at any t, which can be determained from the Dirac propagator function.

89. Furthermore, since *all* objects interact (at least gravitationally), and objects exist only in space and time and *not* in nonspace, it is unclear what the meaning of interaction in nonspace is. That is, since the concept of object is now seen to be essentially a *classical* one, objects do not exist "in" nonspace. Therefore, it makes no sense to say that objects *interact* in nonspace.

90. What is measurement? Measurement, as done by classical objects, is a position coincidence of two objects—implying, of course, that both objects have positions—or the distance or interval between two position coincidences of objects, implying that such a distance exists. But rarely are there position coincidences unless one of the objects is a classical object; i.e., one that is always in space. An object is now called a quantum object if it is only momentarily a classical object. Therefore, I reformulate the concept of measurement: a (classical) measurement is the position coincidence of two objects, or the distance determination between the objects, where at least one of the objects is a classical object (generalized in section 81). Thus, unless an object is a classical object, it rarely has a position and unless there is a continuum of classical objects between two objects, even though they be classical themselves, is there a distance between them.

91. The world, as stated above, is nonspace with restrictions. This nonspace is the seat of all possibilities except for those restrictions (and others, yet to be mentioned) that already have been discussed in this work. What these possibilities are, are simply all possible paths in nonspace with the mass and number of objects restrictions. Space and time, being properties of classical objects, exist only where classical objects exist. Measurements (classical) are defined only in terms of space and time and thus require classical objects. A measurement (ex-

cept in the most exceptional circumstance, when it results in an eigenvalue of the particular kind of measurement made) is (when made on nonspace) a transformation; that is, it is a transformation from one state of nonspace and imaginary time to another, or, less frequently, to space and real time. The latter transformation to space and real time is a restriction on nonspace, but it is not so much of a restriction that it produces a space-time determinism. Classical determinism is an extraordinarily restrictive situation requiring minimum uncertainty, or the classical uncertainty relationship at every instant of time, $(\Delta x)(\Delta p) = \hbar/2$. One way of clearly seeing the distinction between the classical space-time world, a highly restrictive world, and the world of "all" possibilities (nonspace), is that in the former an object can have only one of all possible classical random walks going from one spatial point to another, while in the latter, a single object, not restricted, follows all of these walks together. The restrictions on nonspace that give rise to classical objects, that is, objects in space and time, arise from the nature of at least some objects as interactive (or nonspace to be self-interactive). Where these restrictions, manifesting themselves as interactions arise from, is not known anymore than the restriction required by mass.

92. Because of the nature of nonspace, restrictions on it allow a much greater variety of possibilities than would be true of space-time. Thus, suppose that the nonspace was restricted by a given value of mass and number of objects, say, two. Then, if these were the only restrictions, the state of the system may be defined as a superposition of states, each defined by a different mass, so that if measurements were to be made on this nonspace, the objects being produced would have all possible masses restricted only by the sum of the masses being fixed.

93. The restrictions need not be only mass restrictions. The representation of the state of nonspace as a Hilbert space can lead to an infinity of other kinds of restrictions (a kind of superposition of mass and space-time restrictions). I do not deal with this issue here except to mention it. (Other kinds of restrictions on nonspace are generally implied, such as 3-dimensionality and a Euclidean metric on the distances.) In fact, it

does not seem possible to claim that the number and kinds of restrictions are even finite. In the end, the ontology I have sought turns out to be not the concept of object, but that of nonspace and the restrictions on it; how these restrictions arise, to repeat, is yet to be discovered.

94. Now that we see that nonspace is the fundamental reality (with imaginary time), perhaps the term "nonspace" which is somehow the "negation" of space, indicating that space is the more fundamental reality, and similarly "imaginary time," should be redesignated. I would call the former now *funda* and the latter *fundat* (for "fundamental"). Together, I would call them *fundam*.

XI

Ontology II

95. In this work I have attempted to keep before me the question, "What is reality, what is the source of the results of our measurements?" One result of this investigation is that the concept of reality itself can be analyzed further. Reality has two aspects; actuality and potentiality. Actuality is what is, the state of being of the reality (here, the state of the nonspace), at a given instant of time. Suppose that at some time the nonspace is put into a given state by a classical array of objects. That classical array, in defining the state of the nonspace, defines what I will call the *actuality* of the nonspace. Also at that instant of time, the classical array defines the *potentiality* of the nonspace; i.e., what the possible states of the nonspace are (will be) *if* there is a later time; i.e., if objects will exist. In between these two times (which, of course, exist only if objects exist), the nonspace did not exist *in* time or space, although its state in between these two times must, nevertheless, be describable in terms of space and time. If a measurement is made on the nonspace at the latter time, then the potentiality of the nonspace is transformed into a (new) actuality and a new potentiality arises; that is, the state is transformed into a new state. This occurs even though the measurement may or may not give rise to an object or objects. Note that reality (here, nonspace), that is, actuality and potentiality, are *defined* in terms of space and time; if space and time do not exist, that is, if classical objects do not exist, nonspace, may, nevertheless still exist.

The potentiality of a reality defined at a given time is all future actualities and some non-actualities. That is, the future exists in the present as both actuality and non-actuality (in nonspace and imaginary time, of course); this is the meaning of potentiality. Furthermore, it is through the concept of potentiality that the future is distinguished from the present. Another way of making this point is to say that the present is "wide" enough to embrace both future actualities and non-actualities.

As long as no measurement is made on nonspace, there is determinism; of course, this is not classical determinism, but rather determinism in nonspace and imaginary time. That is, all the future is in the present, no non-actualities exist in the present. Thus, as claimed before, there can be no time in nonspace; only if a measurement is to be made is there even the *potentiality* for the existence of time, space, and object. Furthermore, since within space there is only random preference in the "walk" of an object, there can be no order and, therefore, there can be no time. Time requires order. It is only the existence of the increasing number of positions of the nonspace with time—its increase in spatial extent—that defines the monoticity of time and is evidence for its existence. Thus, not only objects arise from measurements on nonspace (and imaginary time), but also space and time. However, whether time is monotonically increasing or decreasing or is doing actually both cannot be distinguished or determined unless there exists another variable, perhaps such as electric charge. This appears to be why anti-matter in increasing time is identical with matter in decreasing (going backwards in time) time; e.g., positron and electron. Thus, the state of the nonspace is a superposition generally of both forward and backward times, or equivalently, of matter and anti-matter. Such existence is only a potentiality until a measurement is made, which is the transformation of the potentiality into an actuality. That is, what exists in between measurements is not time but only the potentiality of time. That is the meaning of imaginary time.

96. It was shown in the chapter on classical physics that the classical object cannot give rise to a non-contradictory concept of change. If the object is defined at an instant of time, that is, if all its derivatives (an infinity of) are defined at an instant of time, then they are defined everywhere in the analytic domain (of time). That is, the statement of what reality is at a single instant of time defines reality for all time; the restrictions due

to the definitions of the derivatives are so great that no freedom is left for anything but definition; potentiality is reduced to actuality. On the other hand, if the object is incompletely defined so that not all its derivatives are specified, there is no predictability at all; that is, it is an either all or nothing situation, probability not arising. That is, future and past reality either is or is not defined in the present. On the other hand in quantum mechanics, where nonspace is the basic reality, its nature is such that behavior *within* it is deterministic but *probabalistic* at a measurement. That is, in quantum mechanics, as derived here, future reality exists in the present but only as a *potentiality*; a measurement is a transformation, therefore, of a potentiality into an actuality. Change, therefore, is nothing but measurement, where measurement is the transformation of potentiality into actuality and the creation of a new potentiality. *The concept of change therefore makes sense only as a quantum mechanical concept.*

By saying that the world consists only of nonspace or objects (and their interactions) I don't mean to say that there aren't phenomena. There are colors, sounds, etc.; even coincidences can only be made if a coincidence is distinguished from a non-coincidence and this requires the presence of color, sound, sensation, or other phenomena. Eventually though, all phenomena arise from objects in their interaction. Thus, the very nature of measurement, arises, in the last analysis, from nonspace in its "self-interaction." Conversely, the very nature of objects, nonspace, and imaginary time, expresses itself only in its self-transformation into classical objects, and space and time.[12]

This work is now concluded.

Notes

1. Earlier and preliminary versions of the work here can be found in my papers in *Physics Essays: The Non-Structure of Physical Reality*, volume 1, number 3, 1988; and *Quantum Mechanics and the Special Theory of Relativity from a Random Walk*, volume 3, number 1, 1990.

2. L.I. Schiff, *Quantum Mechanics*, 1949 1st.ed., pg.42.

3. An excellent source of the various theories of space and time can found in the two books of Richard Sorabji; *Matter, Space, and Motion*, 1988, and *Time, Creation, and the Continuum*, 1983.

4. I have found one of the most basic treatments of special relativity given to be in V.A. Ugarov, *Special Theory of Relativity*, 1979, published by Mir.

5. W. Feller, *An Introduction to Probability Theory and Its Applications*, volume I, 1950, p.170.

6. The problem of Buridan's Ass is discussed in *The Encyclopedia of Philosophy*, volume 1, p.427.

7. The path integral approach of Feynman and the propagator function for the Dirac equation suggested by him can be found in Feynman and Hibbs, 1965, *Quantum Mechanics and Path Integrals*.

8. I wish to thank V.A. Karmanov of the Lebedev Physical Institute in Moscow for bringing to my attention the works of Gersch, and Jacobson and Schulman mentioned in the text. Furthermore, he proved to me that the Dirac equation could not be derived from a random walk of fixed step length, as I had first thought, and that the procedure of Jacobson and Schulman was correct.

9. The works mentioned in (8) above can be found in: H.A. Gersch, Int. J. Theor. Phys. 20, 491: T. Jacobson and L.E. Schulman, 1984, J. Phys. A: Math Gen. 17.

10. The proof that the state function $\Psi(x,t)$ is derivable from measurements on $|\Psi(x,t)|^2$ and $(\partial/\partial t)|\Psi(x,t)|^2$, was given in 1933 by E. Feenberg (in E.C. Kemble, *The Fundamental Principles of Quantum Mechanics*, Dover Publications, Inc., 1948, pp.71–72).

11. Many years ago as a graduate student I puzzled over how mass could be measured. According to quantum mechanics the acceleration of an object is not defined and thus mass could not be measured, even in a relative manner by Mach's definition. Also it could not be measured by a balance, since both position and velocity values are required, thus contradicting the Uncertainty Principle. Apparently,

in principle, many measurements, as indicated in the text, are required.

12. Only after reaching the ontological site did I realize that Dirac had
already built the foundation and that Feynman had pointed the
way.